DOMAINE LEFLAIVE

PULIGNY-MONTRACHET
DEPUIS 1717

This book is a tribute to life – both the wonders of today's living world and the lives that came before us.

It also celebrates the four generations of Leflaives who, year after year, have shaped the Domaine, and especially its founders, Joseph Leflaive and his wife Camille, our great-grandparents, who set this family adventure in motion.

We dedicate these pages to our children, our grandchildren, and our nephews and nieces, who will have the responsibility and the pleasure of carrying this legacy forward and ensuring the continued existence of the unique blend of dream and reality that Domaine Leflaive embodies today.

Huge thanks to the whole team at Domaine Leflaive, without whom this book would not have been possible.

All images shown in the book reflect the beauty of life and have not been digitally altered.

A TRIBUTE

Domaine Leflaive

PULIGNY-MONTRACHET
SINCE 1717

Edited by
BRICE DE LA MORANDIÈRE

Photographs by
JÉRÔME BRYON

TO LIFE

SKIRA

PREFACE

By Brice de La Morandière

This book is intended as a tribute to life, in all its forms – from the minuscule and microscopic, often hidden beneath our feet, to its fuller expression in our wines and in the emotions they evoke.

In Puligny-Montrachet, we live every moment to the full, as a blessing – from the beauty of the landscape that surrounds us, to the step-by-step development of the vines from dormancy to fruition at harvest. We cherish, as a heritage from past generations, the drystone walls, the stone huts known as *cabottes*, and the mosaic of the historic vineyards that shape Burgundy's vernacular landscape – a reverence that follows us, all the way into the near-spiritual silence of our cellars, where the wine slowly develops and ages.

Working with nature is a reminder that the care given to the smallest detail in the past resurfaces and shapes who we are now. When the flora, in its infinite nature, bursts into life around us, when small animals and even bacterial life regain their place among the vineyards, we realise just how generous nature is when cared for with love. To see nature as a living world is also to accept it as mortal. As a result, we must also recognise our responsibility towards its permanence and towards Humanity.

We live intensely connected to nature and to our roots in the village – every day, every season, every year. We aim only to respectfully contribute, by practising a form of viticulture that cares for the plants, for the terroir and for the diversity of its expressions. The mission of the vigneron is to capture, as fully as possible, the place and the vine in the unique expression of each wine. In a way, it is very much like playing music: the vigneron reveals the beauty of a tune through his or her own interpretation and sensitivity.

The revelation of great terroirs depends not only on the work of the vigneron but also on the contribution of other specialists – from geologists to vine geneticists and oenologists. Just as a whole orchestra is needed to interpret a symphony, we need the wholehearted involvement of each contributor to give the wine its fullest voice. It is only together that we form part of the wider community of life.

This community stretches across time. For us, it embodies the enduring presence of the Leflaive family in Puligny-Montrachet, who first settled there in 1717. The Domaine as we know it today was founded in the early twentieth century, when Joseph Leflaive – a graduate of the École Polytechnique at the age of twenty – chose to share the dividends of his career in industry with his ancestral village. To the 3 hectares held historically by the family in the nineteenth century, he quietly added vineyards and houses between 1910 and 1930, which now form the core of the Domaine as it exists today.

Joseph was forward-looking: each parcel had to be planted – and often replanted – in a Burgundy still recovering from phylloxera and the World Wars. Though the wines had already gained international recognition by the late 1930s, particularly in the post-Prohibition United States, the Domaine struggled to balance its books. Joseph would go on to support it financially throughout his life, knowing that his children would possibly have to do the same, and perhaps quietly hoping that they might one day reveal the full potential and brilliance of Puligny's wines. In tribute to that vision, we continue to reinvest as much as possible in the vineyard and in the Domaine – and, importantly, we remain fiercely independent.

Each generation plays its part in building the family's reputation. The wines we enjoy today would not have this sense of magic without the patient work of four generations of the Leflaive family. Each in turn has contributed to the same shared effort, standing on the shoulders of giants. Resolute, committed characters, drawing strength from their differences, they have shaped that same work. Following their father, Vincent and Jo raised the Domaine to excellence and international renown. Anne-Claude then opened the family's eyes to the importance of preserving the life of the soil and the vine's vigour, so as to pass on the authenticity of our terroirs.

I have had the privilege of taking up the torch for the fourth generation. Rooted in our Burgundian heritage and our family values, I carry the flame without letting the ashes of tradition hold us back – to echo Gustav Mahler's words, which inspire me every day. I feel like a musician – not a soloist but a conductor, in love with a piece of music, with a symphony that can only be performed by an ensemble playing in perfect harmony. The composer's score is that of Puligny-Montrachet and of Burgundy. It has been written over millennia, even millions of years.

The Domaine Leflaive teams enjoy playing its music. We take the greatest care of our instruments: our plants, our vines and the cellars that receive their fruit. Each year, we play this music with passion to convey the emotion to its fullest. Over the course of a century, Domaine Leflaive has become a composition, built layer upon layer, sustained by a deep and passionate bond with the vine and the terroir. Paying tribute to life is as much the acknowledgement of the inherent passage of time as it is the determination to provide continuity for future generations.

Our emotions are all the more beautiful when they come from a true sense of familial loyalty. Each vintage is a newborn that we cherish deeply. Each is the fruit of a year of hard labour and care, built on that of the previous generations, reflecting the union between people and nature that we seek to protect every day.

A web of invisible relationships exists between the land, the soils, the plants, the micro-organisms, the insects, and the animals that come to graze in our vineyards. Who are we, as humans, to stand apart from them? The vine interacts with everything around it, whether vegetal, animal or human, for it teaches us the importance of listening actively to the subtle communication between all forms of life – perhaps imperceptible, but very much present. I humbly admit that we do not understand everything that happens around us. Perhaps that is something we should even welcome: when we accept uncertainties, we become more open and better able to respect the world around us.

Domaine Leflaive is recognised as a pioneer in applying the principles of biodynamics. While we practise it with sincere, daily and rigorous commitment, we do so without dogmatism or esotericism. Biodynamics works at Domaine Leflaive because it is grounded in observation and in the constant questioning of our practices: What benefits the soil, the plant, the wine and our environment? What does not? What can we do better? Together with other vignerons, we practise a form of biodynamics that is both disciplined and pragmatic, driven by a love of life and of the living world.

We must also acknowledge that, in placing too much emphasis on reason or logic, we have at times become somewhat complacent. We are far from knowing it all. Many things that we did not know ten, fifty, one hundred or two hundred years ago are now clear to us. If we cannot demonstrate the scientific benefits of biodynamics now, should we reject it out of hand? Fortunately, scientists themselves embrace doubt, and turn systematically to observation and repeatability before validating their theories. Following Jacky Rigaux, the eminent wine writer and champion of terroir, who offered a compelling defence of biodynamic farming, I am convinced that it is about bringing together scientific knowledge and a respect for the environment and circumstances – a synthesis embodied by two nineteenth-century figures, the chemist Louis Pasteur and the physiologist Claude Bernard. 'The living world exists only through its interaction with the environment,' notes Rigaux. From a philosophical perspective, we might find a similar contrast between Descartes and Kant – between *Discourse on the Method* by René Descartes and *Critique of Pure Reason* by Immanuel Kant. Pure logic overlooks the fact that life brings a purely aesthetic emotion, a capacity to delight in the beauty of the world.

Every day, from my garden in the heart of Puligny, I marvel at the living spectacle of nature bringing life to our vines. This humble village garden is a window on to my world: the world of wine, which has expanded remarkably over the past two decades, and the world of my wife, my three children and myself, who, before finally settling in Puligny, toured the world for thirty years to discover the differences that shape us and bring us closer together. In this garden, you might find a Chinese composition, a Turkish rosace inspired by an Iznik tile, blazing maples reminiscent of autumn in Pennsylvania, or even an English-style square garden bed – reminders of our first encounters abroad. From this vantage point, there is a natural continuum towards the vine-covered slopes, now known throughout the world. The genesis of this book also lies in this garden: in the desire to know and understand both life and the living world.

Before me lies a land that has become almost mythical in the world of great wines: Montrachet, Bâtard-Montrachet, Bienvenues-Bâtard-Montrachet, and a little higher, at the edge of the woods, Chevalier-Montrachet. Two realities present themselves to us – the world beyond, ever-changing; and a timeless Burgundy, deeply rooted for over two millennia.

The vineyard has shaped the Burgundian landscapes. Those we see today are relatively recent, the result of a historical evolution towards single-crop farming. Along the Côte de Beaune and the Côte de Nuits, a narrow strip some 2 kilometres wide and 60 kilometres long is carpeted with vines, which need agroforestry and biodiversity to maintain their balance. Yes, Burgundy is beautiful, but nothing can be taken for granted.

Burgundy must not become opportunistic; it must remain genuine, grounded, generous and, above all, humble. For the magic to endure, we must honour those who came before us, those whose groundwork let us develop the wines we produce today. We must also consider the long-term future of what we have inherited and what we will pass on. To do so, we must place respect for life and for the living world above everything else.

The forager and herbalist Thierry Thévenin reminds us of the extraordinary diversity of the wild plants that surround us, and which have inspired the culinary creations of chef Hugo Bourny. Together, the two men awaken the flavours of this silent world of plants and the life that weaves its way around our hurried steps. It is all right here around us – so close and so beautiful. If, for Wagner, music begins where words end, we can no doubt say the same of the beauty of nature.

Burgundy's landscape has not always been as we experience it today. In the Place du Pasquier de la Fontaine in Puligny-Montrachet, what is now the Domaine was once a farm, and the *climat* Les Houlières, just behind its walls, was once an orchard. The land has become a vineyard; the farm has been repurposed. It is our responsibility to ensure that these transformations respect what came before. We converted an old bottling space into a reception room, making use of the existing beams and stone. Likewise, in the historical *cuvage* on Rue de l'Église, we erected a contemporary facility in cut stone, using a local limestone called Beaunotte, one of the mineral treasures that make up Burgundy's subsoil. The living world is present in our workspaces too, something we must always keep in mind. The wines must flourish in symbiosis with their wider environment.

Fortunately, the people of Burgundy know how to appreciate and feel connected to their heritage – by keeping it alive in their homes and villages and by recognising it collectively. Looking after our heritage in this way is increasingly the role of the leading figures of Burgundy's vineyards and of institutions such as the association Les Climats du Vignoble de Bourgogne. With the support of international patrons and donors, the association helps to restore drystone walls and renovate *cabottes* – the small drystone shelters that have long stood among the vines. We do this work so they may be used again, not turned into lifeless shrines to the past. Above all, Burgundy must not become a museum piece!

Our ambition is not to preserve, but to adapt, to enhance, to elevate. We make wines that are not exactly those of previous generations, because circumstances, our environment, our knowledge and our experiences have changed over the generations. What connects us is expressed in these characterful wines, crafted for longevity, and which, as I see it today, are like the world itself – in motion. We interpret them each time a little differently, and yet we continue, each day, to honour what they have been.

What binds us to one another and to nature gives rise to unparalleled emotions. This is the full artistry of wine, the poetry it evokes: capturing within a bottle the full expression of a place and of the people who call it home. This is our Domaine Leflaive. This is what this book invites us to do – to encounter something greater than ourselves. Life and the living world.

Page 9
Puligny-Montrachet village

Left
Brice de La Morandière

Following spread
Ammonite, a silent witness to Burgundy's geological history

THE SKIN OF THE EARTH

At the top of the Chevalier-Montrachet plot, high on the slope, a 150-million-year-old past is laid bare along the face of the embankment. Like a slice of history, two distinct layers of rock emerge, exposing sedimentary strata from the depths of time. One is compact and solid, made of limestone; the other is a soft and crumbly bed of clay-rich marl. The two layers alternate, piling up like a millefeuille that has absorbed all the earth's tremors. Here, in this small corner of the Puligny-Montrachet vineyard – home to some of the world's finest grands crus – there is no need for specialist knowledge to grasp its subtleties. Even the casual walker can observe the partly unexplored world in the soil beneath their feet.

Surprising as it may now seem, what is now Burgundy once lay in the shallow depths of the sea. Its waters teemed with life, with many organisms dying and settling into sediment as the waters gradually receded. Today, their fossils remain – shells, in particular. Few are visible on the surface: they erode quickly. Yet just beneath lies a hidden treasure, revealed by a light brush of the hand. The teams at Domaine Leflaive sometimes uncover remarkable specimens, such as an ammonite with its perfectly coiled shell preserved in stone. It is a trace of evolution from another era entirely, and yet it is still the same earth.

Page 15
Small tortoiseshell butterfly

Left
Geological strata at the top of Chevalier-Montrachet

Right
Cabotte, *meurger*, drystone wall and vineyards at the Dos d'Âne

In our age of mapping and modernity, science has given names to the rocks that have shaped this landscape. There is the limestone of Chassagne, lower down on the celebrated Montrachet slope, whose geology is especially rich. Ladoix limestone is its close cousin. They take their names from villages in the Côte de Beaune, and they too bear this history in the stone of their walls. Over time, humankind has put down roots in these ancient soils, occupying and shaping the land. Today, the grapes grow in a landscape marked by human activity. Yet it remains one of the most natural expressions born of the earth and one of the most beautiful, when left free of artifice. It is a testament to what humankind can achieve in harmony with nature.

Left
Puligny-Montrachet vineyards

Right
The mare Prune at Nantoux

Page 21
Biodynamic compost

The words 'humus' and 'human' share a common Latin root, which simply means 'earth'. If the intimate bond between humankind and humus has been broken, how can we relearn to live alongside soils whose sustainability is vital? How can we reconnect with these living paths that we literally and figuratively trample beneath our feet? We tamp them down, compressing and compacting them underfoot, at times rendering them barren. We neglect them too, in a more philosophical sense; we have unlearned their importance.

Marc-André Selosse, a biologist at the Muséum National d'Histoire Naturelle in Paris, continues to drill home the simple truth: 'Soil is the origin of the world.' When healthy, it sets in motion a virtuous cycle, fostering mycorrhizae – the vital connections between roots and fungi. Soil also retains water, prevents erosion, captures carbon from the atmosphere, and much more.

Left
Cypress-leaved plaitmoss

Right
Winter stalkball

The vigneron plays a part in preserving the natural environment, restoring it, bringing vitality back to the soil, and encouraging the return of essential organic matter such as micro-organisms and the humble earthworm. The future of generations to come is at stake.

Domaine Leflaive takes an active role in caring for the land and for the wider environment that surrounds it. The vineyard has developed a pioneering ethical approach over several decades, rooted in organic farming and biodynamics. Free from ideology or esotericism, we apply these rigorous methods humbly in our everyday work. They help to strengthen the vines, protect the local flora and fauna, and respect the earth's natural rhythms.

This ethical approach is a form of biodynamics rooted in working *with* the land. It calls for a simple, committed effort by humankind to move towards a natural harmony. Even if we cannot fully explain all its benefits, why turn away from it when its effects are clear? This grounded approach is a far cry from celestial magic or shamanistic rituals. It is a warm tribute to every dimension of life, as Brice de La Morandière observes: 'We're rediscovering a natural connection with the soil.'

This daily approach forms the backbone of Domaine Leflaive. It is a framework, a set of practices that need no explanation. 'It's something that runs through each of us naturally; it comes up in conversation,' we say as we read the vines to determine their needs, checking the colour of the leaves and the vigour of each young plant. Winemakers develop an instinctive feel for the plant, maintaining a connection with the soil and contact with the living world.

We plant legumes, brassicas and grasses as cover crops between the rows of vines. These species and more play their part in aerating the soil, adding organic matter, and gently filtering water into the ground. For years, we have used horses for ploughing at the Domaine. The horse's soft tread avoids compaction and encourages microbial life, breathing life into the soil. The soil breathes.

This vital energy gives the wine its character, shaped by its home, its 'terroir' – where people and place come together in productive dialogue, not a dogma or a checklist of slopes and exposures. A terroir is made of subtle energies that rise from deep within the earth, flowing through the winemakers, becoming wine. And the finest wines are born of living soils, of a profound respect for the earth.

From left to right
Ladybird, round-headed
garlic and wild garlic

PORTRAIT

FRANÇOISE VANNIER

Geologist

Françoise Vannier knows every parcel of Domaine Leflaive, its soils and, above all, the subsoil beneath – the very ground that the winemakers walk and whose secrets, properties and composition they seek to unlock. Françoise is a geologist. She brings clarity to these complexities through a simple methodology rooted in observation. 'I know much more now than when I started in 2000,' she says with a smile. 'In geology, experience counts for a lot.'

Technology also plays its part, with electromagnetic waves probing the depths and soil pits dug to complement the surveys, among other methods. Over several months, she studied Domaine Leflaive's holdings in Puligny-Montrachet, Mâcon and Nantoux. Crucially, she studied everything in depth, digging beneath the rocks and identifying the soil composition. 'I realised that Burgundy's soils had never been studied with viticulture in mind. And yet no one *climat* is the same. Even the term "clay-limestone" has many nuances within a single *climat*.'

Now settled in the region, Françoise has worked across vineyards in Sancerre, Champagne and Corsica, and further afield in Menorca, Spain, Chile and the Sta. Rita Hills in California – even as far as Hokkaidō in Japan. Yet her travels always lead her back to Burgundy.

PORTRAIT

CAROLINE BESSE

Decorative artist and painter

In 2022 the decorative artist Caroline Besse was awarded the Grand Prix de la Création de la Ville de Paris for fine art craftsmanship. She quickly caught the attention of Brice de La Morandière for her subtle work using natural materials. She is one of the very few artists in the world to paint using minerals ground to a fine powder, on large format, thanks to a technique she developed herself.

After meeting Brice in Burgundy, Caroline set out to capture the relationship between the soils of the Domaine's vineyards and the organic expression of its wines. It all began with a collection of mineral samples from the Puligny-Montrachet parcels. Caroline crushed the minerals to create pigments and prepared plant decoctions to make inks. This first stage enabled her to create a map of the Domaine using its own natural colours, initially expressed through test tubes containing the raw materials themselves and then through colour charts presenting a palette of the Domaine's soils. 'I then used the colours to create visual imagery inspired by tasting the Domaine's cuvées, translating the sensations I felt in the moment into graphic forms.' In doing so, Caroline Besse confirmed a deeply held conviction: 'Minerals are the building blocks of all matter: they shape the soil, nourish the vines and make up our own bodies.'

‘TO BRING AGRICULTURE AND NATURE INTO HARMONY, WE MUST BEGIN BY LOOKING BENEATH OUR FEET.’

Gaspard Kœnig

Left
Black mustard

Below, right
Vine clippings used to create a dead hedge

Left
Yarrow in winter

Page 46
From left to right
Thierry Thévenin,
Hugo Bourny and Brice
de La Morandière

DOMAINE LEFLAIVE

GASTRONOMIC CREATIONS IN TRIBUTE TO THE TERROIR OF PULIGNY-MONTRACHET

Brice de La Morandière invited Thierry Thévenin and Hugo Bourny to walk the land of Domaine Leflaive over the course of a year.

Thierry Thévenin is a botanist, forager and leading voice for herbalism in France. He compiled a list of plant species found on the estate, whose diversity is enhanced by its commitment to biodynamic practices.

Hugo Bourny, Michelin-starred chef at Lucas Carton from 2021 to 2026, was captivated by the experience: 'These grasses, flowers, shrubs and trees are the quiet witnesses to a relationship between plants and people that is at once precious and fragile, as seen at Domaine Leflaive. It has deeply inspired me ...'

From this shared exploration, the idea of creating a series of unique dishes quickly emerged, bringing together the life, the landscapes of Burgundy and the Domaine's wines through exceptional pairings.

THE SILENT KINGDOM OF PLANTS

By Thierry Thévenin

KEEPERS AND SENTINELS OF TIME

The wild plants of Domaine Leflaive reflect land stewardship practices committed to life as a whole. They are the expression of a vineyard cultivated biodynamically for over thirty years on an exceptional terroir formed by a long, multi-layered history and a range of influences.

As steadfast witnesses and active participants in evolution, they carry memory and signals of change. They speak of times gone by and hint at possible futures for those who take the time to observe them.

May the pages that follow honour this remarkable legacy. May these few plants, and their equally remarkable light, help brighten our present at a time when the future feels uncertain.

MYSTERIES OF ORIGIN, FORCES OF PLACE

Mountain inula (*Inula montana*), European pasqueflower (*Pulsatilla vulgaris*) and the exceedingly rare goldilocks aster (*Aster lynosiris*) tell the story of Palaeolithic herds of chamois roaming the steep, open grasslands of Montrachet and Nantoux, long before hunter-gatherers settled and became shepherds and cultivators. Typical of dry, sun-soaked grasslands, and silent witnesses to those ancient times, these species were long sustained by grazing flocks of sheep until the early twentieth century. But as pastoral life declined, the species lost ground,[1] and forest and vine advanced up the slopes.

The king of the forest here is downy oak (*Quercus pubescens*), the tree that gave its name to Quercy as the 'land of oaks'. It thrives in these parts because, like the vine, it prefers limestone soils, rocky ground and hot summers, all of which lie at the heart of the legend of Burgundy's great wines.

Like any sovereign, the oak in all its majesty does not stand alone. Common boxwood, gathered for centuries and blessed in church on Palm Sunday, is fading from the landscape as the box tree moth has taken hold. Yet wild madder (*Rubia peregrina*) root and barberry (*Berberis vulgaris*) bark can still yield abundant reds and yellows, as they once did for the textile dyers of southern France.

While flora from the south is still present here, with flowering ash (*Fraxinus ornus*) and Montpellier maple (*Acer monspessulanum*) at the very northern edge of their natural range, the presence of the mountains is felt just as much.

On a clear day, the distant dome of Mont Blanc rises on the horizon from the Chevalier-Montrachet vineyard. It is no surprise, then, to find that same grandeur and strength in its wines – nor to encounter a few mountain plants such as Alpine buckthorn (*Rhamnus alpina*) in the scrub along the Montrachet ridge, or narrow-leaved red valerian (*Centranthus angustifolius*) among the scree below the Dos d'Âne.

Between the Seine and the Saône, beneath the Alps and the Jura, within reach of the Mediterranean's sea breeze, the Côte de Bourgogne forms a meeting ground where its *climats* come into their own.

1. Today these species are protected in the Burgundy region, most notably *Aster linosyris*, which is on the brink of extinction there, threatened as scrub moves in and the vine spreads. It is undoubtedly the most precious heritage species found at the Domaine.

2. What botanists would call commensal plants of the vine.

VINEYARD FLORA (*FLORA VINEA*)

The vineyard flora is made up of companion plants[2] that thrive alongside the vines themselves, benefiting from the meticulous, routine work of the growers. Year after year, these species mark the turning of the seasons, nourishing the life of the soil and protecting it from erosion.

From autumn through to the first pass of the hoe in spring, the vines are carpeted with what we call 'vine salads': garden cress, dandelion, plantain, burnet, shepherd's purse, hawksbeard, meadow salsify and sow-thistle. These all germinate after summer storms and spend the colder months at ground level, their central buds nestled deep within rosettes of leaves, protected from hard frost and grazing animals. They form a living skin over the soil, shielding it from winter rains. When worked back into the soil during the first hoeing in March and April, earthworms transform them into valuable organic matter.

In conventionally managed vineyards, species including grape hyacinths (*Muscari neglectum* and *Muscari comosum*) and wild garlic (*Allium vineale*) are now in decline, or even absent. At Domaine Leflaive, shallow hoeing and the decision not to use herbicides mean these plants can thrive.

In summer they are protected from the burning sun and drying winds by their underground bulbs, formed of fleshy leaves layered tightly on top of one another. This strategy, native to Mediterranean regions, evolved in response to the recurrent wildfires typical of this climate. The bulb remains sheltered from flames and embers; the plant sends up its leaves, flowers and seeds in the winter and spring, when fires are rare.

Wild garlic, like cultivated garlic and all species of the *Allium* genus, releases sulphur compounds into the soil, which help limit the spread of vine pathogens such as downy mildew and powdery mildew. It is not a 'weed' but a true ally of the winegrower.

From this perspective, there is no such thing as weeds. Calling them such is to ignore their message and their virtues. Coltsfoot, well established on the lower slopes of the Mâconnais, thrives in unstable soils that are highly prone to erosion. Its dense network of long, interwoven rhizomes acts like a living bandage that binds the soil and preserves its structure. It encourages the winegrower to work the soil lightly and less frequently, preferring the tread of a horse and its handler to the weight of a tractor.

Wild plants can even act as early warning signs, helping to predict harmful conditions. Field bindweed, henbit dead-nettle and cutleaf geranium all suggest that the soil is not holding on to nutrients well, allowing excess nitrates to leach away, and that the soil is beginning to compact. When an attentive winegrower notices these wild plants, they know it is time to step back; otherwise, the vine will suffer, weakened by excessive nutrients and poor aeration.

A VALUABLE AND FRAGILE RESOURCE

Wild plants are a treasure we need to preserve, not least for the care they offer to the soil, which in turn benefits people and animals alike.

Their benefits are often manifold, as seen in yarrow (*Achillea millefolium*), nettle (*Urtica dioica*), dandelion (*Taraxacum officinale*) and valerian (*Valeriana officinalis*), which readily grow in the *chaintre*[3] – the headland, an unworked strip of land at the end of each row where horses and machinery turn.

These four plants provide essential ingredients for biodynamic preparations that help the soil absorb and regulate nutrients, strengthen the vine's resistance to disease and pathogenic fungi, and, in the case of valerian, temper the effects of late light frosts.[4]

The same four plants have long been valued across Europe as essential medicinal remedies. In Burgundy, yarrow was used for the liver and for regulating menstruation, while nettle was used to treat diarrhoea in people and animals. Dandelion was used for a wide range of everyday ailments. Its flowers were made into syrups for bronchial conditions, its roots taken as a decoction, and its leaves taken as an infusion for the stomach and eaten in late-winter salads for their cleansing properties. Across France, valerian root was used for anxiety and sleeplessness, and was also thought to ease the tremors of old age. The list of these traditional remedies is long. Many have since been validated by scientific research and adopted by modern medicine.

Mountain inula (*Inula montana*) stood in for arnica for shepherds in Puligny and Nantoux. It was used to treat bumps and bruises when it still grew abundantly on the hillsides. Infusions and tinctures of field marigold (*Calendula arvensis*) are highly effective when used as compresses to heal minor wounds; marjoram (*Origanum vulgare*), mallow (*Malva sylvestris*) and wild thyme (*Thymus praecox*) aid digestion and soothe the lungs; and ash leaves and couch grass root support the kidneys and the urinary tract.

Even the vine itself has its role. Its leaves, when applied with honey, help bring boils to a head, while its sap – collected in small bottles at pruning time – was until recently used as an eye lotion for treating conjunctivitis and other eye conditions.

3. The *chaintre* is also known in some regions as a *tournière* or a *tavela*. Far from being wasted space, it provides a refuge for a wide range of plant and animal species, many of which play a valuable supporting role in the vineyard, while also offering a variety of resources, from small game and snails to medicinal plants and mushrooms.

4. 'Les préparations biodynamiques : un bien commun' [Biodynamic preparations: a common good] ; https://www.bio-dynamie.org/pratiquer-la-biodynamie/preparations-biodynamiques/bien-

EMISSARIES OF
LOBALISED WORLD

Plants are often the secret passengers of cross-continental travel in today's globalised world. From Puligny-Montrachet to Solutré, Domaine Leflaive has, in this sense, also become a vineyard of the wider world, as seen in the presence of Japanese knotweed (*Reynoutria japonica*), narrow-leaved ragwort (*Senecio inaequidens*) and staghorn sumac (*Rhus typhina*) beneath the Dos d'Âne, and Guernsey fleabane (*Erigeron sumatrensis*) in the Clos des Pucelles.

How did this small patch of Japanese knotweed come to be established along the lower slopes of Les Perrières, at the Champ Canet vineyards in Puligny, and is it liable to take over, in keeping with its reputation as an aggressive invader? Japanese knotweed is unlikely to spread unless its surroundings are disturbed by major earthworks – the very conditions in which it thrives. Might its young, tender shoots one day find their way into the kitchen in a spring dish, as they do in Japan?

In Japan, the plant is known as *itadori-kon*, loosely translated as 'root of well-being', and is valued as an essential remedy in traditional medicine. Its rhizomes are used to treat liver disorders, ease joint pain, reduce inflammation and infections caused by bacteria and fungi, and help with delayed menstruation and chesty coughs. Its powdered dried roots and leaves are also used in *kanpō*[5] medicine to help heal wounds and burns, and, when applied locally as decoctions or infusions, to relieve allergies. It contains catechin and quercetin, which support these long-standing uses. Recent research also points to high levels of resveratrol, a recognised anti-inflammatory and antioxidant linked to cardiovascular protection.

At Domaine Leflaive, Japanese knotweed is not considered an enemy but rather an ally, as its infusion can help protect the vine against mildew. Narrow-leaved ragwort lies discreetly beneath the Dos d'Âne, while staghorn sumac makes a striking impression in autumn, its foliage turning a flaming red, just like its velvety clusters of fruit. In North America, Indigenous peoples used it to make a refreshing, tangy drink akin to lemonade. They also used various species of fleabane for their analgesic and anti-inflammatory properties. Fleabane is also known to be effective in treating pain, including lumbago and tendonitis, as well as urinary infections and kidney disorders.

These species most likely arrived at the Domaine over the course of the twentieth century and have since become part of its flora, thriving amid the many ecological and climatic disturbances brought about over the past two centuries. They may well foreshadow the landscapes of the future and hold the promise of valuable resources for generations to come.

PORTRAIT

THIERRY THÉVENIN

Botanist, forager, author

HONOURING THE MAGIC OF THE SOIL

‘My first encounter with Domaine Leflaive was a voyage of discovery, setting out to explore plants destined to become flavours, textures and aromas in Hugo Bourny’s cuisine, and colours and lights in Jérôme Bryon’s artworks. The journey took me to realms between vineyards and forests, north and south, stone and soil, where human hands have spent centuries lovingly tilling, tending, planting, weeding and pruning, nurturing the land with care to coax forth wines of unparalleled excellence.

‘I returned season after season with grateful delight, treading the first fresh-ploughed furrows in the morning mist, foraging in the divine light of dawn beneath the Dos d’Âne and sharing in the great shimmering sun of the harvest. I walked the vineyards and thorny footpaths of Montrachet with joy in my heart. I strode on the Nantoux hillside and the serried rows of red earth in Solutré, on a quest for the tiny treasures that form the superb, age-old, living heritage of Domaine Leflaive. I returned time and time again in recognition of the promise to keep spinning the magic of superb wines by honouring the magic of the soil and living in harmony with birds and insects, grass and stones.’

Left
Coltsfoot

Right
Round-headed garlic
Birdeye speedwell

‘BEING MOVED IS LIKE
BREATHING WITH
YOUR HEART.’

Pierre Reverdy

Page 61
White stonecrop

From left to right
Vineyard peach tree,
blackthorn, juniper

Left
Lamb's lettuce

Right: from left to right, top to bottom
Yellow head spurge, amaranth, ribwort plantain, Austrian pine, barberry and yellow mustard

Left: from left to right, top to bottom
Fumitory, upright bugle, pasqueflower

Yellow mustard, wild carrot, dandelion

Pasqueflower, narrow-leaved red valerian, bulbous meadowgrass

Right
Saint Anthony's turnip

THE HUM OF WILDLIFE

For many of us, a sense of wonderment at all forms of life around us no longer comes easily, as we have gradually lost our connection to nature. For a long time, winegrowers kept wildlife at bay, believing it to be harmful to the vines. But we do not need to go feral to reconnect with wildlife or to see it teeming all around us. Where the land is treated with respect, the soil becomes richer and begins to host a multitude of minute species, such as worms, insects and mites – all valuable allies that help maintain the balance of the ecosystem.

In the vineyards of Domaine Leflaive in Puligny-Montrachet, the lacewing, charmingly nicknamed in French the *demoiselle aux yeux d'or* ('golden-eyed damsel'), is a minute but effective regulator of pest populations. Various species of tit can also be spotted in the Houlières vineyards near the village, and hoopoes, with their long beaks and warm orange plumage, return each year to nest at the top of the Chevalier-Montrachet grand cru plots. They all have a role to play in maintaining a healthy ecosystem. On the heights of Nantoux, where the Domaine has planted a parcel of Chardonnay, green lizards and their larger, legless cousin, the slow worm – often mistaken for a snake – can sometimes be seen.

Right
Fox tracks

The magic of the living world preserves life in all its forms – from the smallest at our feet, the unseen beneath our steps, to the larger creatures around us. The signs are everywhere, though often fleeting. To notice them, we must hunker down to the eyeline of smaller animals, whose far keener senses catch the slightest trace of a wild presence.

It might be the tracks of a wild boar in snow, or the fleeting shadow of a roe deer pausing to drink. Subtler still, and tinged with a touch of mystery, are the signs the trees themselves yield up: bark worn smooth where animals have rubbed against it, a branch a bird has nibbled in passing. On the ground, pine cones gnawed by a squirrel or a vole are much easier to spot. You might even be lucky enough to find a pair of antlers shed by a stag at the end of winter.

Left
Robin

Right
Roe deer hair caught on a branch
A trail used by roe deer, wild boar and foxes

Following spread
Starlings in flight

Around the vineyards of the Côte de Beaune, the landscape is naturally less hospitable to larger creatures, which tend to seek the shelter of woodland far from human activity. At the top of Chevalier-Montrachet, however, a few wild boars sometimes slip through in the early morning. In the Beaune region, hares, badgers, foxes and roe deer roam the edge of the woods, coming close to the vines. Around the Montagne de Cras, above the village of Nantoux, the long-standing practice of extensive grazing has preserved soils that attract different species.

Today, pockets of biodiversity help create natural corridors for wildlife. Partial grass cover between the rows, drystone walls and low hedgerows can provide suitable habitats for birds such as robins, nightingales and swallows. Birds of prey, including peregrine falcons, circle overhead. 'Burgundy needs to foster biodiversity to maintain balance across all forms of life,' says Brice de La Morandière.

Eco-grazing is a natural extension of the Domaine's philosophy. It strengthens the bond between humans and animals, even within the modest confines of a wine estate. The benefits of eco-grazing are difficult to measure in practical terms alone. We know it reduces the need for mechanical weeding, but its true significance goes beyond that. The practice reflects a deeper relationship with the land, rooted in care, attention and respect. Through eco-grazing, connections are forged between animal, soil, plant, shepherd and winegrower. A natural harmony takes shape – perhaps the wine itself even retains some trace of it.

When sheep first began grazing among the vines in Solutré at Domaine Leflaive, they quickly learned how to move between the rows. No doubt animal instinct plays its part, but so too does the ability of the flock to work independently and to adapt to its surroundings. Working together, the shepherd and vineyard manager made observations, exchanged ideas and refined their approach. Today, the sheep graze throughout the Domaine's vineyards in the Mâconnais and in Puligny-Montrachet. These remarkable four-legged gardeners know exactly how to nibble beneath the vines without damaging the vine stocks. They help keep winter ground cover in check as it grows increasingly vigorous in a changing climate, naturally fertilising the soil as they graze.

Bringing animals back into the vineyard also helps restore ecological balance. Plant life regenerates, biodiversity flourishes and birds return to an environment in which they can thrive. During the nearly two months they spend among the vines, the sheep enjoy a welcome supplement to their diet. For the shepherd, the arrangement provides valuable support for a livelihood that remains an important part of the rural economy. For the Domaine, the simplicity of the practice is in keeping with an approach that favours restraint over intervention.

The same spirit guides the use of horses for ploughing. Far from being a matter of folklore or appearance, traditional ploughing offers clear agronomic advantages. The horse's softer tread keeps the soil light and allows it to breathe. Roots reach deeper, organic matter builds, the soil absorbs heavy rainfall more easily, and erosion is less likely. The practice also helps reduce the vineyard's carbon footprint by replacing some heavy tractor work.

Right
Common blue butterfly

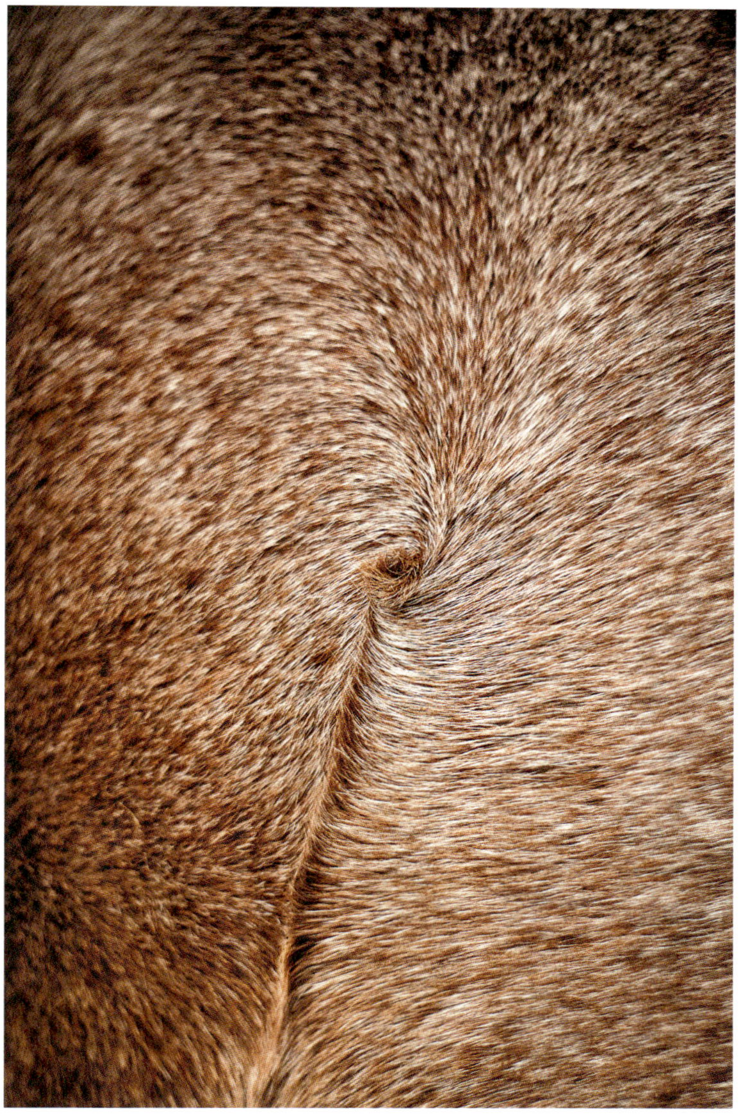

The soil calls for a light touch. Animals instinctively know this, and tread gently on the earth. Research by Marc-André Selosse, a biologist at the Muséum National d'Histoire Naturelle in Paris and author of *The Origin of the World*, has shown how intensive mechanisation over recent decades has degraded soils, stripping them of their vital force and in some cases even rendering them barren.

Domaine Leflaive is one of the few estates that keep their own horses. The mares Prune and Lola worked the land at Puligny-Montrachet for many years before enjoying their well-earned retirement in the meadows of Nantoux. Now, ten-year-old Gitan has picked up the baton. He is an Auxois, a Burgundy draught breed trained for vineyard work. To complement this effort, the estate also brings in horses from elsewhere. Ginette and Image, two mares from the Beaujolais, work alongside Gitan. They are brought in several times a year to till the steep slopes of the Mâconnais. The mares can work up to six hours a day; guided by voice and by the friendly barks of Maloue, the vineyard dog, they skilfully manoeuvre through the delicate areas around the old vines.

The mere presence of the horse is a real boon and a return to nature. When soil, animals and people come together, they form a simple expression of life. This fruitful exchange makes the horse not merely a tool but a true companion to the winegrower, a living link in the complex workflow that shapes the wines.

To cultivate the entire vineyard in this way would, of course, require a stable of horses. At the Domaine, we meet this challenge through a system of rotation. We use horses for one year, the next year we use machines, and the third, we leave the vineyard untouched.

Our approach is also guided by the character of each vintage: we work the soil less in hotter years to conserve moisture and we check grass growth in wetter conditions so the vines can thrive. When we walk the land ourselves, we immediately feel that working with the horses is the right approach. The soft, supple soil beneath our feet tells us so.

Left
Gitan, the horse of Domaine Leflaive

PORTRAIT

JULIEN MORVAN

Shepherd

The shepherd and sheep breeder Julien Morvan named his venture Tandem Animaux in honour of his 200-strong flock of sheep on his farm in Magny-Jobert, in Haute-Saône. He works in protected natural areas managed by regional conservation bodies and for private businesses, grazing his sheep on their green spaces as an alternative to lawnmowing. 'But it isn't just about trying to appear "green" for the cameras,' he stresses. Julien is attentive to animal welfare and uses pastoral farming methods to support biodiversity. He often observes birds newly nesting in the grass. The 'tandem' also refers to the bond between his flock and the places where they graze, renewing an age-old tradition as he moves between summer and winter pasturelands. At Domaine Leflaive, he began with 2 hectares in Mâcon and now works 10, in addition to 6 hectares in Puligny-Montrachet. In step with the rhythm of the vine, four groups of twenty-five sheep graze across the different patches over a period of fifty days, from early February to the end of March. 'I'm there all the time,' he says. 'It's difficult to delegate because you need to know your animals well. Tending an estate requires skill and sensitivity, with a good measure of intuition.'

PORTRAIT

PHILIPPE RODRIGUEZ

Traditional ploughman

Philippe says lovingly of his two mares, 'It's important they enjoy their work so that they want to come back when we fetch them from the field. It's a relationship built on trust and exchange, where the horse gives rather than us taking from it.' Formerly a shepherd, Philippe Rodriguez now works with animal traction, which is increasingly sought after by wine estates for parcels that are difficult to farm with machinery. Domaine Leflaive calls on him for its vines in Solutré. Philippe calls out in a lively tone to guide Ginette and Image, the two mares he raised with care. Image is an Auxois from northern Burgundy, and Ginette an Ardennais, two of the roughly ten draught breeds still found in France. They are eight years old and will work the soil until around twenty before retiring to pasture in Fleurie, in the Beaujolais. Over the years, Philippe has come to understand the full range of their mettle, whether in agricultural work or hauling timber for the forestry trade. Many training schools have sprung up recently, and the equipment used has evolved too. The old ploughs of the pre-tractor era are firmly in the past. Harness makers and farriers continue to refine the tools of the trade, making them even more ergonomic. It is a movement very much alive and in demand.

A LIVING DIALOGUE

HUGO BOURNY

‘For me, Domaine Leflaive turned out to be a playground where my cuisine naturally sparked up a dialogue with the vineyards and the surrounding landscape. I set out to develop an approach that honoured the four seasons – nothing showy, just quiet confidence. I imagined my recipes in response to the terroir, the soil, the climate, and the labour of the men and women who work in the vineyards, seeking to echo that raw emotion and capture the depth of a wine, the light falling aslant on a hillside, the memories rooted deep in a grape variety. I think of it as telling a story that begins amid the vines. I learned discipline and demanding standards from the masters, but they also taught me to listen to the moment. At Domaine Leflaive, this tradition has its own special meaning: each dish pays homage to nature, translating the gorgeous landscapes of Puligny-Montrachet into flavours that speak the language of Burgundy. My cuisine plays with the local flora like the memory of a fragrance, the bright petals of a flower or the power of the land. Wines are my creative partners, inspiring me with their honesty and intensity. I consider it an intuitive conversation that celebrates the bond between we humans and the living world – a quest for truth, depth and resonance.’

Hugo Bourny’s career has been shaped by a quiet commitment to the highest of standards. He trained for over fifteen years with some of the world’s leading chefs, including Arnaud Donckele and Anne-Sophie Pic. 2026 marked his fifth year as head chef at Lucas Carton – a crucial milestone that has seen him update the finest Parisian tradition with a contemporary eye for a cuisine whose watchwords are emotion, intuition and precision.

SEASONAL GREEN ASPARAGUS WITH COMMON YARROW

Hugo's creative pairing notes for Brice

The crunch of asparagus spears tells us spring is here. Yarrow adds notes of camphor, summer savoury, white florals and a hint of bitterness.

A fine vegetarian dish with a touch of acidity offset by a yoghurt sauce whose dairy note and kick of voatsiperifery pepper create an impressive depth of flavour.

The tender yarrow stems add an elegantly green bitterness, while the vinaigrette highlights the star of the show, asparagus, with a delightfully sharp, acid edge.

Mâcon-Verzé Le Monté 2021

This asparagus dish calls for a young, lively, enthusiastic wine such as the Mâcon-Verzé Le Monté 2021 to harmonise with the subtle bitterness of the asparagus and acidity of the vinaigrette. It needs a wine of some complexity to hold its own with the green notes, especially the yarrow, which plays a major role in biodynamic wine production.

Mâcon-Verzé
LE MONTÉ
LEFLAIVE - PULIGNY

MARINATED RAINBOW TROUT WITH RIBWORT PLANTAIN

Hugo's creative pairing notes for Brice

The rich flavour of the marinated – but raw – trout is accentuated by the slightly numbing, clove-like, citrusy and tangy Sichuan pepper.

The ribwort plantain broth is tangy like a vinaigrette, cleansing the palate with its fresh, green plant notes.

The camelina oil adds notes of fresh lettuce heart to match the stem lettuce.

Bienvenues-Bâtard-Montrachet Grand Cru 2018

It takes a majestic wine with plenty of discreet grace to honour trout without overpowering it. A Bienvenues-Bâtard-Montrachet, particularly a young vintage like this 2018, will whisper with the fish and sing with the broth, starring the ribwort plantain found everywhere in Puligny vineyards.

Bienvenues-Bâtard-Montra
GRAND CRU
DOMAINE LEFLAIVE - PULIGNY-MONTRA

FIELD TOMATOES MARINATED IN WILD FENNEL

Hugo's creative pairing notes for Brice

Marinated, lightly seared, raw tomato hearts give this dish plenty of punch and all the warmth of a tomato plucked fresh from the vine.

The fennel shoots bring a touch of liquorice, while the fennel flowers add a lasting aniseed kick, and the Citrus Pursha introduces the sweet-and-sour notes.

The smoked fromage blanc ice cream softens the overall effect.

Meursault Premier Cru Sous le Dos d'Âne 2014

The delicate, subtle 2014 vintage is the perfect partner for the fennel-liquorice notes and hint of acidity in this dish.

Meursault 1er Cru
SOUS LE DOS D'ANE
DOMAINE LEFLAIVE - PULIGNY-MONTRACHET

SLOW-COOKED PEAS AND WILD MARJORAM

Hugo's creative pairing notes for Brice

The peas are slow-cooked in a pea pod jus, then blended to create a fresh green taste that captures their natural sweetness.

The camphor-eucalyptus note of silver fir is a perfectly matched foil for the wild marjoram, while the other herbs and plants highlight the diversity of plant notes for the palate.

Puligny-Montrachet Premier Cru Les Folatières 1992

This recipe calls for the finesse and maturity of a beautifully evolved vintage from a vineyard renowned for its subtle balance. After thirty years, the honey and camphor notes of this Folatières 1992 are ample reward for the connoisseur's patience.

Montrachet
LES FOLATIÈRES

VENISON LOIN SMOKED WITH HARTWORT ROOT

Hugo's creative pairing notes for Brice

The hartwort root tempers the gamey aroma with its smoky, medicinal depth – earthy and faintly acrid.

The barberries add a confident, bright tang that cuts through the roundness of the meat. The Swiss chard is served almost raw, briefly sweated in a dab of warm butter.

The sauce is deep, powerful and well seasoned, with a strong kick of pepper from the red long peppercorns.

Bâtard-Montrachet Grand Cru 2019

The majestic, full-bodied character of this grand cru, renowned for its power, makes for a rare, perfect marriage between game and white wine. The 2019 vintage offers notes of linden, hazelnut and dried fruits. The exceptionally long finish shows off the unusual pairing to its finest extent.

Bâtard-Montrach
GRAND CRU
DOMAINE LEFLAIVE - PULIGNY

YOUNG CARROTS, FROM LEAF TO SEED

Hugo's creative pairing notes for Brice

A dish blending sweet and acid savours.
The carrots retain their crunch, while the greens bring a fresh, grassy aspect.

The peel praline creates an interesting texture; the beads of lemon gel are a fine palate cleanser.

But the soul of this dish is the sauce: a malted reduction offers depth and a pleasing bitterness that forms a striking contrast with the carrots.

Serve with a sabayon flavoured with marc de Bourgogne, sweet vernal grass, tonka bean, carrot seed and brown butter.

Pouilly-Fuissé Premier Cru En Vigneraie 2020

This vintage saw the Pouilly-Fuissé family join the wine world elite, earning the well-deserved title of premier cru. The generous, concentrated mouthfeel of this Pouilly-Fuissé Premier Cru En Vigneraie is redolent of citrus, white florals and a hint of almond – aromas that underscore its sheer elegance.

Pouilly-Fuissé 1er Cru
EN VIGNERAIE
DOMAINES LEFLAIVE - PULIGNY-MONTRACHET

CANDIED, CONCENTRATED BEETROOT AND JUNIPER TWIG

Hugo's creative pairing notes for Brice

Extremely slow-cooked beetroot, candied to remove the earthiness while retaining a succulent bitterness brightened by a fresh beetroot, butter and juniper vinegar glaze.

Served on a tangy bed of sorrel, and diced beetroot in lime, with a rich beetroot, sumac, butter and gin sauce.

The juniper twig lends the sauce a unique plant note that is greener and paradoxically less resinous than the more familiar juniper berry.

Puligny-Montrachet Premier Cru Les Pucelles 2015

Delicate, elegant and subtle, Les Pucelles reveals itself to those who know how to wait. Selected here in a rich vintage, it offers wines of excellent concentration and marvellous salinity.

LES PUCELLES
DOMAINE LEFLAIVE - PULIGNY-MONTRA

LINE-CAUGHT WHITING, ROSEHIP VELVET AND BURGUNDY TRUFFLE

Hugo's creative pairing notes for Brice

This recipe was built around the signature sauce's deep, tangy and velvety notes. The rosehips bring a bold, bright vivacity, with notes of redcurrant – smooth and invigorating.

The assertive heart of the dish is the lightly cooked fish, its purity of flavour intact – a perfect ally for the nobility of Burgundy truffle.

A dish worthy of contemplation and a true homage to winter, a season of contrasts.

Chevalier-Montrachet Grand Cru 2017

It might seem iconoclastic to pair whiting and truffle, but the resulting marriage demands the grandiose Chevalier-Montrachet 2017 – one of the greatest vintages of recent decades. A wine of this distinction calls for a touch of reverence: doff your hat and bend a knee as you raise your glass!

Chevalier-Montrachet
GRAND CRU

SYLVAIN GOUJON

Pastry chef

Sylvain Goujon is head pastry chef at the world-famous Lucas Carton restaurant. He is renowned for his precise, pared-back creativity. Working closely with Hugo Bourny, he conjures up desserts that align perfectly with the restaurant's culinary vision. His hallmark is drastically slashing the quantity of sugar in his desserts to let the ingredients shine.

He excels at developing bold new pairings, often using fresh herbs and spices to showcase a carefully controlled acidity. His visuals pay due homage to the history of their setting, with a strikingly modern twist. He is an outstanding master of his art, playing with textures to round off a meal at Lucas Carton with a light yet memorable flourish.

PINK GRAPEFRUIT, POLYPODY FERN AND SAGE

A fantastic showcase for acid, bitter and cleansing notes. Polypody root offers a woody, delicately sweet liquorice flavour.

The dessert merges a slight kick of woody pepper with the liquorice notes of polypody root.

Pink grapefruit is the star of the show, ushering in a delightfully low-key acid bitterness, while sage adds a quiet hint of camphor.

The result is a heady blend of crisp, complex savours.

Fine de Bourgogne

Fine de Bourgogne will amplify the dessert's aromatic profile. It is aged in oak barrels for a decade, giving it a vivid, crystalline freshness borne by ripe citrus notes and hints of wood and liquorice, finishing with spicy, slightly camphor notes that echo the dessert's own sage and pepper. The overall impression is one of a harmonious blend of vivacity and aromatic complexity, with a deeply elegant finish.

MAINE LEFLAIVE

PORTRAIT

HENRY
DE MONCLIN

Designer of bottles and decorative objects

This project was Henry de Monclin's first foray into the world of wine – but unlikely to be his last, given his deep affinity for Burgundy. Henry is a design consultant specialising in the perfume sector. He divides his time between Paris and New York, creating perfume bottles for leading fragrance brands (from Jean Patou to Valentino) and for music artists (such as Beyoncé, Rihanna and Katy Perry), each seeking to express their identity through the medium of scent. 'Each project reveals something new about others and about myself,' he says. 'It's a world for aesthetes – you need a keen eye and the ability to work without preconceptions'

When Brice de La Morandière approached him in January 2025 to create a bottle for a fine de Bourgogne from Domaine Leflaive, the challenge stood out in an otherwise standardised field. After several months of reflection, the design took shape. It began with a 35 centilitre carafe, its soft, fluid form composed of stylised eggshell shapes – a subtle reference to the Domaine's Egg Cellar (La Cave de l'Œuf). The carafe is produced by Waltersperger, recognised for its excellence in craftsmanship and one of the few glassmakers where traditional blowing and pressing techniques remain central. This expertise, rooted in the Bresle Valley in Seine-Maritime, a region renowned for luxury glassmaking, is echoed in the striking bottle stopper. 'Brice and I exchange ideas constantly,' Henry explains. 'We moved away from our initial idea of using wood. The inspiration came from the stone of the walls surrounding the Chevalier-Montrachet vineyards.' The stone is crushed, cast with resin in a mould, then polished by the craftsmen at MMB VOLUM-e to create a stopper like no other.

HANDS OF THE HARVEST

Burgundy likes to say that these lands were destined for wine. Its unique geology, its generous soils, its balanced climate that shapes each vintage – all make it the perfect partner for grape vines. Burgundy winegrowing has come to embody a form of terroir-driven viticulture recognised worldwide, where the essential work is done by sun and soil. And yet, human beings are part of this miracle that is born anew each year. Without human insight, no place can truly come into its own. The vineyards of Burgundy reveal a harmony between nature and the human hand, which plays its part with modesty. An invisible thread binds the two. A quiet current runs through the daily work, raising a simple question: What is the right way to approach the land, one that lets life and the living world express themselves? And how do we harness its generosity to coax forth great wines, while respecting the land and its needs over time, never taking more than it can give?

Having converted to organic farming, Domaine Leflaive helped pioneer the biodynamic approach in Burgundy some thirty years ago. Since 2015, Brice de La Morandière has continued this commitment, not as a form of magical thinking that removes the need for human care, but as a vision for the future in which people still play an essential role. The vine is guided and supported rather than left entirely to its own devices. As Amandine Brillanceau, technical director and head of the vineyard teams, explains: 'We're all deeply attuned to biodynamics. It gives meaning to what we do and connects us to nature. We accept that some vintages are more challenging. Our resilience comes from the tools we use, but also from the experience of the people and their memory of the land.' Work in the vineyard is carried out with sensitivity and restraint, with a quiet sense of purpose and, at times, a touch of grace. Everything is done with calm assurance. As the author Jacky Rigaux notes: 'Winemakers committed to biodynamics uphold a form of viticulture that respects biodiversity and remains as close as possible to the natural balances of the most renowned vineyard sites.'

Pages 152–153
Puligny-Montrachet
village

Left
Amandine Brillanceau, technical director at Domaine Leflaive

Following spread
October vineyards, whose autumn colours inspired the name Côte d'Or

‘HISTORIC BURGUNDY RESEMBLES A POINTILLIST TAPESTRY.’

Bernard Pivot

Right
Julien Coyaud, vineyard manager at Domaine Leflaive

Cultural practices here are guided by conviction and by choice, and call for both discipline and a strong sense of purpose. When the Domaine acquired pastureland with clear potential for Chardonnay in Nantoux, in the Hautes-Côtes de Beaune, it invested several years of careful preparation before any vines were planted. The team took time to observe how the soils responded, working to preserve the existing ecosystem while encouraging biodiversity. Legumes and cereals were sown to nourish the land, and the soil was worked lightly by horse rather than by machine. Only at the end of this lengthy process did the wines slowly begin to reveal their personality, inviting a patient and attentive eye. Faster methods might have been possible, but the Hautes-Côtes de Beaune deserve the same care as the grands crus.

Caring for the vines is a collective effort that resists a rigid approach, instead responding to the needs of each stretch of land. In keeping with the Burgundy tradition of working parcel by parcel, pruners, tractor drivers and vineyard workers spend long periods in the *lieux-dits* of Les Houlières or La Plante des Champs. The attention they pay to each vine grows out of a close, daily familiarity. 'Staying connected to the plant means everyone has to be versatile,' explains Julien Coyaud, vineyard manager in Puligny-Montrachet, as he walks the soils of Les Brelances, heavy in the late autumn. 'The hardest part of our work is the unpredictability of nature, but we accept it.'

Across the Mâconnais, in the villages of Verzé, Solutré or Saint-Véran, cultural practices are guided by the same pursuit of understanding and excellence. The dialogue with nature remains measured: it calls for exploring and testing, then adopting only techniques that have clearly proved their worth. Cover crops, for instance, were trialled over time before being introduced more widely.

On the more recently acquired Jobeline estate, Laurent Alexandre has led the conversion to organic and biodynamic farming with energy and conviction. He has identified distinctive parcels and closely followed the development of the orchard of apple, cherry and vineyard peach trees near the winery, monitoring its contribution to biodiversity. Here, experience comes before dogma, restraint before excess. Established practices are not taken for granted. Even pruning – an age-old practice – continues to evolve through observation, shared knowledge and the skill of those who carry it out.

Right
Laurent Alexandre, vineyard manager

'WINE IS EARTH'S ANSWER TO THE SUN.'

Margaret Fuller

In the wild, grape vines grow untrammelled, following the sun and the breeze. For winegrowing, they must be gently shaped and guided by human hands, but never contorted or forced. This calls for a sensitive, painstaking touch.

The Domaine has therefore chosen to follow training methods developed by the specialists at Simonit&Sirch. Their work is guided by a single objective: preserving the vine's vitality. The aim is to let the vine grow a little each year – a technique they describe as controlled branching – rather than holding it back, as is more commonly done. This knowledge must be relearned, adapted to each individual vine, and passed on. The health and longevity of the vineyard, and the ability of the vines to age harmoniously and produce great wines, depend on a close, attentive sense of kinship with the plant.

Left
Baptiste Guigon,
tractor driver

Right
Vine shoots

MARCO SIMONIT

Co-founder of Simonit&Sirch, master vine pruner

Few people understand time as Marco does. The son of farmers from Friuli, Marco Simonit spent nearly twenty years patiently developing his approach to pruning. The first vines planted by his grandfather for the family's wine fascinated him. 'I always wondered what was happening inside a vine. No one ever spoke about it. One day, I cut open a vine that was failing but not quite dead, and saw a line of white wood that was still alive.' That moment of discovery, rooted in careful observation, marked a turning point. With the support of a professor at the University of Milan, he was able to refine his ideas, followed by a first commission that confirmed the value of his method. Between 2005 and 2006, the renowned Piedmontese vigneron Angelo Gaja invited him to train his teams of pruners. Working alongside his partner, Pierpaolo Sirch, Marco Simonit soon found the doors of the world's great vineyards opening to him. In France the great vigneron Denis Dubourdieu introduced him first to Bordeaux, then to Burgundy.

At Domaine Leflaive, Simonit&Sirch now work across the entire vineyard, from Puligny-Montrachet to the Mâconnais and, more recently, Nantoux, where they were involved from the earliest stages of planting and training. Their approach is always bespoke, tailored to the terroir, often unfolding over several seasons and rooted in the philosophy of lifelong learning. For Marco, *savoir être* lies at the heart of *savoir faire*: 'You have to live and breathe a vineyard to understand it – its terroir, its history, its practices. Exchanges with winemakers are essential; it's a relationship that is constantly evolving.' Full of ideas, Marco has found a deeply committed team at Domaine Leflaive. 'We work together to ensure the health of the vines. Because to produce site-specific wines, you need living wood in the vines.'

CAROLE DUMONT

Master vine pruner

Burgundy has a way of drawing you back. An agronomist by training, Carole Dumont spent seventeen years in Chile advising wine estates before returning to France. In the fast-moving world of global wine, she met Italians Marco Simonit and Pierpaolo Sirch in South America. The dynamic pair have reshaped approaches to pruning since the mid-1990s. Back in Burgundy, she has represented Simonit&Sirch in the region's vineyards since 2018.

At Domaine Leflaive, her regular visits follow the rhythm of the vine and extend across the entire estate. Understanding, testing and applying the right approach takes time: it often takes several seasons before her painstaking work bears fruit, quite literally. 'Poor pruning practices are a major cause of vine decline,' she explains. 'Here, I've found a highly attentive and committed team.' Together, year after year, they shape what Simonit&Sirch describe as 'the vineyard's architecture'.

GRAPE HARVESTS

Harvesting the grapes at just the right moment is an age-old ritual, bringing people together around a task that is simple yet crucial. The timing of the harvest has become increasingly uncertain as weather patterns change – part of both the challenge and the beauty of making wine. When nature tests our ability to adapt, decisions must be made without ever compromising quality. The harvest calls for a well-practised team.

In Puligny-Montrachet, the forty or so pickers include many familiar faces each year, and the mix of personalities is part of what makes the harvest so special. Harvesters are housed on site and throughout the village, coming together for communal meal-times in the large marquee set up in the courtyard, where a delicious fresh menu is served daily. Young and old work side by side. An eighty-year-old veteran of the harvest rubs shoulders with French workers from the South-West, the North and Brittany. Some pickers come from as far afield as Italy and Spain. And no harvest would be complete without the traditional *paulée* – the celebratory banquet that marks the close of the harvest.

Left
Julie Boisselet,
administrative and
financial director

Right
Bastien Muzard,
head of maintenance

19

In the cellar, Louis Brière tends the wines day by day. Time is his greatest ally, showing him when to intervene gently and when a small adjustment is needed to refine the wines. The Combe winery is also a testing ground, with a range of vessels – stainless steel and concrete tanks, large wooden *foudre* vats, wine eggs and globes. 'The wine is alive and constantly evolving; we cannot be content to settle on a single method,' says Louis. 'But at Domaine Leflaive, the approach to ageing does not follow fleeting trends. Our aim is simply to preserve quality.'

Certain principles remain unchanged, including a year of ageing in oak barrels followed by six to nine months in stainless steel tanks. This second stage allows the wines, first shaped in oak, to come together in larger vessels and express themselves more fully. When the vintage calls for it, extended ageing of up to two years brings out wines of natural refinement and precision. Louis Brière intervenes as little as possible. A great wine cannot be replicated; it changes from year to year, and it is this that gives it both its authenticity and its character.

Regular group tastings are an integral part of the team's work, offering a constant lesson in humility. They help to answer the questions that arise: How are the wines evolving in tank, and which require particular care? How does each wine reflect its terroir, and how does Chardonnay express itself from one site to another? A single grape variety can reveal the character of each vineyard site, further shaped by the massal selection used at the Domaine with its oldest vines. The diversity of individual plants preserves genetic variation and strengthens the vineyard's potential.

Together, the wines of Domaine Leflaive form a family, their varied personalities shaped by the singularity of each terroir and the unpredictability of the vintages. In Burgundy, their magic reveals itself over time, as the wines gradually blossom and release their aromas. It is the role of great estates to present them at this moment of maturity. The life of the vine unfolds over decades, the life of the wine over years. The Domaine's wine library preserves these bottles behind bare stone walls that let the cellar breathe in tandem with the wine. A breath of life and soul.

Right: from left to right, top to bottom
Justine Belloc,
wine scientist
Louis Brière,
cellar master at
Domaine Leflaive
Esteban Pavesi,
cellar master

‘BURGUNDY WINE [...]
REQUIRES A SKILLED HAND TO GUIDE
IT WITH ART AND PATIENCE.’

Jules Lavalle

LANDSCAPES AND HERITAGE: ECHOES OF THE PAST, PROMISES OF THE FUTURE

Winegrowing Burgundy is both singular and diverse. It unfolds as a mosaic of vineyards and a chain of villages, each telling part of a shared story. To speak of Burgundy's fragmented landscape as a coherent whole is more than poetic licence, more than a carefully sustained paradox upheld by its most learned commentators. It is a cultural construction that can be read in the landscape as much as in the written record. Over the centuries, monasteries and abbeys – Cluny, Cîteaux, Saint-Vivant – have helped shape the horizons of Burgundy as we know them today. With time on their side, the monks worked with the future in mind.

Today, the *clos* remains the most visible expression of this heritage; its fundamental unit is the *climat* – a precisely defined parcel of land enclosed by walls. A less visible inheritance lies in a shared way of thinking, passed down through generations. At its heart is the belief that wine emerges from an intimate harmony between soil, site and origin, expressed through more than a thousand *climats*, each offering its own unique expression. Beyond this lies a deeper sense of the singularity of the land itself, conveyed through an atmosphere, a form of wisdom, an unspoken language – a belief in a Burgundy greater than even the greatest of its wines.

In the Côte de Beaune, at the foot of Montrachet, the eye is drawn to one of the most prestigious grands crus, revered by wine lovers around the world and enduring through the centuries. Just as striking is the low stone wall that borders it, a lasting trace of Burgundy's cultural heritage. Many of these walls have disappeared over time, dismantled by changing practices and the demands of mechanisation, which favour easier access. Yet the stones remain. Known locally as *laves*, they are not volcanic but fragments of limestone brought to the surface by rain. The thinner slabs are used for tiles, the thicker ones to build the walls, which can still be seen throughout the vineyards of the Côte de Beaune.

Domaine Leflaive has undertaken the restoration of the wall at Montrachet – not for aesthetic reasons, nor as a museum piece, but because it serves a purpose, helping to manage water and wind within the parcel. It is also an act of faith. Why else restore the *cabotte* at Nantoux, in a vineyard only recently planted in the Hautes-Côtes de Beaune? The presence of this drystone hut, once used to shelter the grower and his tools, suggests that vines once grew here, likely until the nineteenth century, before phylloxera ravaged the vineyard. Like the *meurgers* – piles of stones cleared from the soil during work in the vines – the *cabotte* forms part of the region's identity. It calls for care, but not reverence; it should be preserved without being turned into a shrine.

In 2015 the *climats*, terroirs of Burgundy, were inscribed on the UNESCO World Heritage List as a cultural landscape, recognising a combined work of nature and human endeavour of outstanding universal value. The inscription also underscores the importance of the vineyard's architectural heritage, and the winegrower's duty to protect and preserve it.

Right
Julien Coyaud, vineyard manager at Domaine Leflaive
Gabriel Faquet and Martin Muriot, drystone masons

With the support of its major patrons, including Domaine Leflaive, the association Les Climats du Vignoble de Bourgogne works to restore the region's vernacular heritage, most often in dry stone. It has already supported several hundred projects, led by both private and public owners, all linked to the winegrowing heritage of the UNESCO-listed site. Backed by figures such as Brice de La Morandière and other leading estate managers in the Côte d'Or, the contribution of these patrons has proved decisive. This long-term work also includes training artisans in traditional restoration techniques and supporting the continuation of their craft.

Low stone walls, *meurgers*, *cabottes* and *clos* gates shape the contours of the vineyards, while storehouses, cellars and presses mark the villages. The pale beige hues of Burgundy limestone form part of the region's distinctive visual signature. In Puligny-Montrachet, fragments of history reveal how rural and parish life have intertwined since the eleventh century. The church, entrusted to the abbey of Cluny as early as 1095, and the barn of Blagny, once belonging to the Cistercian abbey of Maizières, are lasting traces of this past. Today the village lives in the secular spirit of its age, even as the whole world venerates it for its great wines.

Left
The Saint-Pierre church spire, Solutré

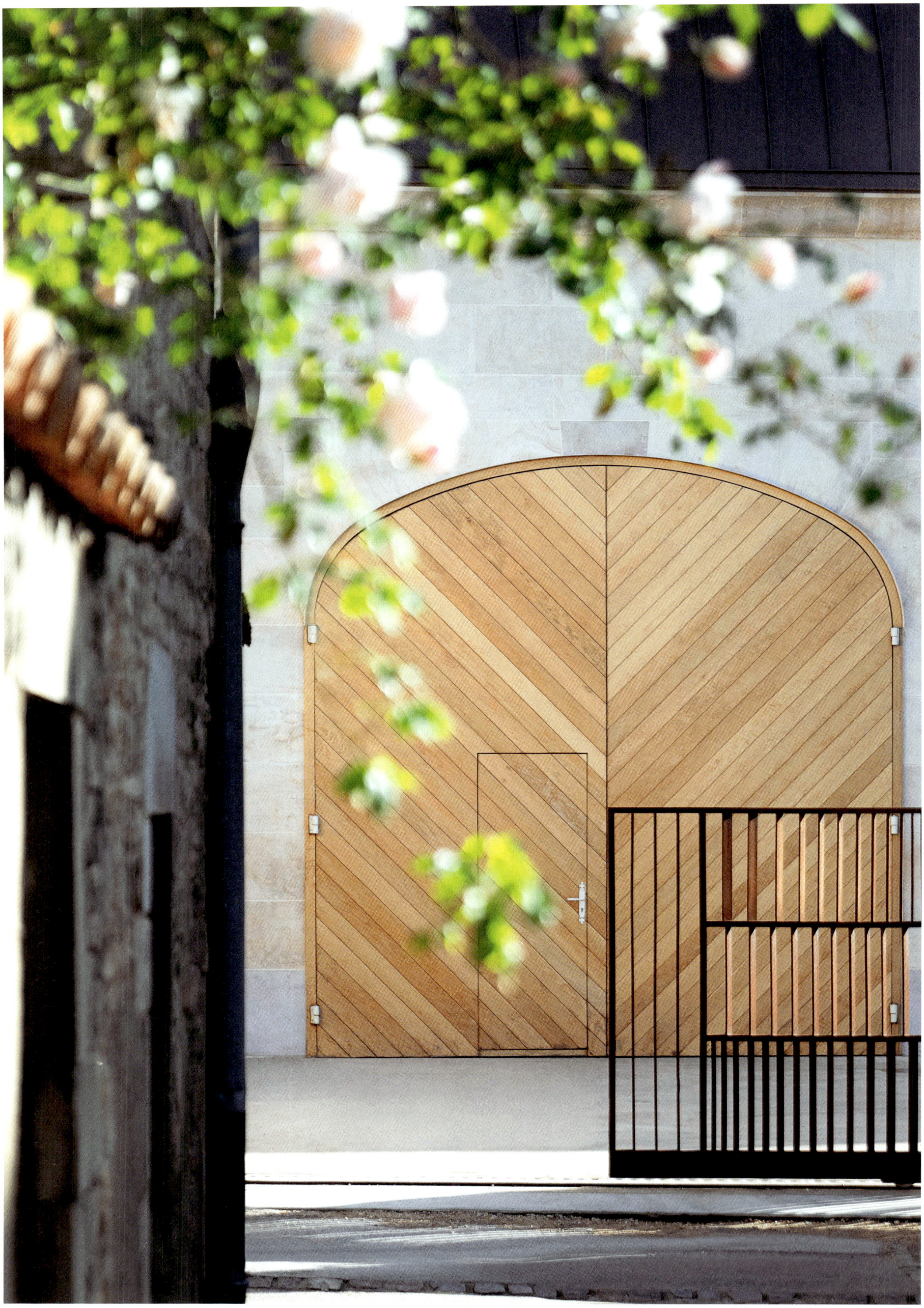

Left
Main entrance
to the winery on
Rue de l'Église,
Puligny-Montrachet

Right
Wood and Beaunotte
stone, detail

Joseph Leflaive, founder of Domaine Leflaive, built the estate from the 3 hectares of vines owned by his family in the nineteenth century, gradually acquiring land and buildings in Puligny-Montrachet. His descendants have never considered moving from the heart of the village, such is their visceral attachment to it. The wines are first the expression of a soil, and then of the vineyard site that nurtures them. They require an environment that supports their development to maturity and beyond.

A building, too, can be seen as a living thing, resonating with the organic matter it shelters and with the people who work within it. It is in this spirit that the Domaine undertook the complete renovation of its facilities. The winegrowing village of Puligny-Montrachet has a distinctive feature – a water table so high that it is not possible to dig underground cellars. As a result, houses are entered from the upper floors, while the ground level is reserved for vinification and ageing. The Egg Cellar (La Cave de l'Œuf), built in 2012 within the winery on Rue de l'Église, reflects both innovation and fidelity to the Domaine's philosophy. Entirely above ground, its egg-shaped form

Left
La Combe cellar

Right
Domaine Leflaive wine library

obeys the golden ratio, evoking life itself. Its timber structure and straw insulation make it fully bioclimatic. The winery on Rue de l'Église has also been renovated to highlight traditional architectural forms, such as the basket-handle vault in the grand cru cellar. Two new buildings have been added in contemporary Beaunotte stone, creating bold interiors and discreet exteriors. Here, substance comes before style, and the architecture serves the wines.

The gate that opens on to the courtyard is made of solid steel over 6 metres wide, with concealed welds and angled bars that give movement to the structure. It is a fine example of craftsmanship in a demanding material, created by the father-and-son duo David and Kilian Villeneuve. David founded the metalworking workshop in Chagny; his son Kilian, named Best Apprentice of France in metalwork, joined him after six years of training. Both are local artisans.

Right
Wine cellar on
Rue de l'Église in
Puligny-Montrachet

DOMAINE

Whether renovating existing buildings or building anew, the aim is always to use resources wisely. When rebuilding, new work must remain in harmony with the extant structures, with materials reused wherever possible.

The restoration of the reception area at Place du Pasquier de la Fontaine reflects this approach. The existing structure was renovated using a circular, resource-conscious method, favouring careful, considered workmanship over purely cosmetic fixes. The walls of the reception room are finished in raw-earth plaster made from soil taken from the Chevalier-Montrachet vineyards. The space remains closely connected to the Domaine's vineyard and its ancient slopes, carrying a deep sense of place. The buildings retain both a physical and emotional link to the land.

Right
The Egg Cellar

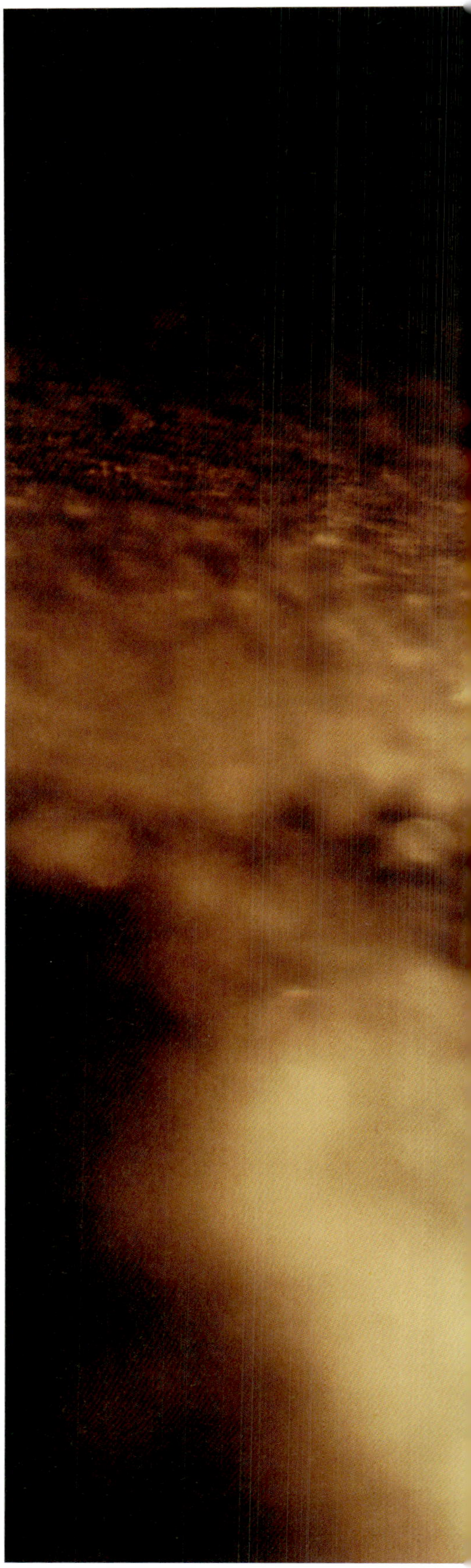

Following spread
Yellow wall lichen

1989

PORTRAITS

MARTIN MURIOT & GABRIEL FAQUET

Drystone masons

Drystone masons Martin Muriot and Gabriel Faquet are bringing new life to an ancient craft now experiencing a revival in parts of France. Rooted in Burgundy, they work with a long-term vision, restoring walls, roofs and shelters using traditional drystone techniques, without mortar. In doing so, they honour the work of the generations of *muraillers* who came before them.

They recently restored a *cabotte* on the estate in Nantoux, at the edge of recently planted parcels. The small stone hut was overgrown with vegetation, its sharp lines softened by a fine layer of moss. It was dismantled, the stones sorted and grouped, then rebuilt piece by piece. Some were reshaped – mainly at the corners – to ensure the structure stood true. Martin raises one finger in the air and traces the graceful curve he pictures for the *cabotte*'s entrance, awaiting the moment when the final coping stone will be gently slotted in place. 'A simple arrangement of stones can become a form of art,' he says with a modest smile.

Page 195
Gabriel Faquet and Martin Muriot, drystone masons, with Julien Coyaud, vineyard manager at Domaine Leflaive

PORTRAITS

ANNA SAINT PIERRE & VICTORIA SOSOLIČ

Artist-designer
& architect

Anna Saint Pierre holds a doctorate in design, following a degree in textile design from the École Nationale des Arts Décoratifs. The questions that shape her work are philosophical in nature, yet firmly grounded in practice: What memory do we retain of a place that is fading? Should it be forgotten, preserved or transformed? She explores how the built environment and its remnants can take on new, decorative forms – bricks becoming flooring, rubble becoming pigment.

Right
The cut stones salvaged from the estate's demolished buildings were carefully preserved, sliced into slabs to follow their natural contours, and re-laid in the courtyard to form a composition in their own right – a quiet testament to their former lives.

Following page
Puligny-Montrachet Premier Cru Les Pucelles, 1978, the oldest bottle held in the estate's cellars

Pages 216–217
Place du Pasquier de la Fontaine

Victoria Sosolič is an architect and founder of Élès Architectures. Her work is guided by an unobtrusive, holistic approach, bringing heritage buildings into contemporary use through local bio- and geo-sourced materials.

For the Domaine, the two women formed a partnership. Commissioned to transform a former bottling facility on the historic site in Puligny-Montrachet into a reception space, they developed a shared vision that brings together art and architecture. The question was how best to approach this transformation: What could be reused on site so that the building remains alive and retains traces of its past? They began by identifying what could be preserved, recycled or repurposed, starting with architectural elements in plain sight, such as the courtyard paving, made from stones recovered from the demolition of two adjacent houses, and extending to less visible details, such as the joints between the stones, made with reclaimed aggregates. These elements, both visible and hidden, form part of a memory that continues to inhabit the place.

Puligny-Montrachet
LES PUCELLES
APPELLATION CONTROLÉE
DOMAINE LEFLAIVE

'WINE IS SOCIAL – IT IS NOT
MEANT TO BE DRUNK ALONE.
WINE IS ABOUT SHARING.
IT IS CIVILISATION.'

Mario Vargas Llosa

8 RECIPES
BY HUGO BOURNY
CHEF

SPRING

SEASONAL GREEN ASPARAGUS WITH COMMON YARROW

Yarrow marinade, charred lemon gel, voatsiperifery pepper yoghurt and a yarrow flower vinaigrette

SERVES 8

ASPARAGUS MARINADE
16 asparagus spears
50 g coarse sea salt
25 g yarrow flowers

FERMENTED ASPARAGUS
4 asparagus spears
2 g sea salt

LEMON GEL
100 ml lemon juice
1.5 g agar-agar
Fine sea salt

YARROW COULIS
100 g yarrow leaves
20 g watercress
Vegetable stock, as needed
3 g lemon gel
Fine sea salt

YOGHURT SAUCE
200 g plain yoghurt
20 ml olive oil
3 g voatsiperifery pepper

YARROW VINAIGRETTE
20 g yarrow coulis
20 ml olive oil
5 g lemon gel
Fine sea salt

TO SERVE
Unsalted butter
Tender yarrow stems
Yarrow flowers
Olive oil

ASPARAGUS MARINADE
Blend the yarrow with a little of the salt, then stir in the remainder. Peel the asparagus spears, trimming off the woody ends, then bed them in the salt to marinate. Check every 30 minutes or so. Once the spears begin to soften, remove them from the salt and rinse thoroughly.

FERMENTED ASPARAGUS
Bed the asparagus flat in the salt in a thoroughly disinfected glass jar. Close the lid tightly.
Leave to ferment at room temperature for at least 1 week. When ready, chop the spears into short batons.

LEMON GEL
Bring all ingredients to the boil and set aside to cool. Once the gel has set, strain through a sieve and transfer to a pipette.

YARROW COULIS
Choose the best bright-green leaves. Strip away the harder stems and blanch for 1 minute in well-salted boiling water. Add the watercress leaves. After 10 seconds, place both in a bowl of ice-cold water, then drain.
Blend thoroughly with vegetable stock and the lemon gel to make a smooth, fairly liquid coulis. Filter and transfer to a pipette.

YOGHURT SAUCE
Blend yoghurt and olive oil. Add grinded voatsiperifery pepper into a fine powder and season to taste.

YARROW VINAIGRETTE
Blend all ingredients, season to taste and set aside.

TO SERVE
Sweat the asparagus spears in a lidded pan over a low heat in barely foaming butter.
Quarter the spears and add a drizzle of olive oil.
Arrange all the elements decoratively on plates, serving the vinaigrette on the side.

MARINATED RAINBOW TROUT WITH RIBWORT PLANTAIN

Savoury ribwort plantain broth
and camelina oil

SERVES 8

MARINATED TROUT

2 trout, 300–500 g each
50 g coarse sea salt
3 g Sichuan pepper, courtesy of Valérie de La Morandière's garden
2 g minced lemon balm

JULIENNE OF STEM LETTUCE

1 stem lettuce

MEYER LEMON COULIS

120 ml Meyer lemon juice
Zest of ½ Meyer lemon
2 g agar-agar

RIBWORT PLANTAIN BROTH

80 g ribwort plantain leaves
80 g stem lettuce leaves
Ginger, garlic, water, as needed

RIBWORT PLANTAIN VINAIGRETTE

50 g ribwort plantain broth
25 ml fish stock (or water)
5 ml Meyer lemon juice
20 ml camelina oil
Fine sea salt

SICHUAN PEPPER YOGHURT

1 plain yoghurt
Sichuan pepper, courtesy of Valérie de La Morandière's garden
Fine sea salt

TO SERVE

Tender ribwort plantain shoots
Ribwort plantain flowers
Tender popping cress shoots

MARINATED TROUT

Dress, fillet and bone the trout the day before. Lay the fillets in the salt with Sichuan pepper and lemon balm for 8 minutes, then rinse. Dry thoroughly before smoking for 10 minutes. Keep the smoked fillets in the refrigerator overnight.
Remove the skin, dice the fillets, and sprinkle with ground Sichuan pepper.

JULIENNE OF STEM LETTUCE

Peel and julienne the stem lettuce heart.

MEYER LEMON COULIS

Bring all ingredients to the boil together for 30 seconds. Set aside to cool until solid, blend, and transfer to a pipette.

RIBWORT PLANTAIN BROTH

Remove the thick stems, then blanch the ribwort plantain and stem lettuce in salt water for 1 minute.
Strain and place in ice-cold water. Strain again, then blend thoroughly with garlic and grated root ginger to taste, loosening with a little water if necessary. Strain the broth through a fine-mesh sieve.

RIBWORT PLANTAIN VINAIGRETTE

Blend the ingredients and season to taste.

SICHUAN PEPPER YOGHURT

Stir the ground Sichuan pepper through the yoghurt.

TO SERVE

Arrange all elements decoratively on a plate, adding the vinaigrette at the very last moment.

FIELD TOMATOES MARINATED IN WILD FENNEL

Candied late-harvested Citrus Pursha, fennel flowers, olive oil, smoked fromage blanc ice cream

SERVES 10

ICE CREAM
15 g sugar
10 g glucose syrup
350 g fromage blanc
3 g fine sea salt
3 g dried fennel stalk
For smoking: 3 g beechwood and 3 g dried fennel stalk

CANDIED CITRUS PURSHA
2 late-harvested Citrus Pursha
20 g blossom honey
20 ml vinegar
15 ml water
Zest of 1 regular lemon
1 g ginger
5 g fennel flowers

SHORTCRUST PASTRY
100 g lightly salted butter
50 g finely shaved parmesan
110 g flour
2 g salt
20 g whole egg
20 g egg yolk
1 egg yolk

TOMATOES
5 pineapple tomatoes
Candied Pursha peel (reserved)
Fine sea salt

TO SERVE
Candied Citrus Pursha jus and paste
Fennel flowers
Fennel shoots
Shortcrust pastry
Fruity olive oil
Fleur de sel

ICE CREAM
Stir the fennel stalks through the fromage blanc and leave to infuse for 8 hours. Strain, then smoke the infused fromage blanc for 10 minutes with a blend of dried fennel and beechwood.
Heat the remaining ingredients in a pan. Remove from the heat, add the fromage blanc and mix thoroughly. Transfer to the Pacojet beaker. Freeze until solid, then churn.
Alternatively, loosen the base with a little milk and serve as a sauce for this starter.

CANDIED CITRUS PURSHA
Cut the late-harvested fruit in half, and separate the pulp. Set the peel aside.
Caramelise the honey and deglaze with vinegar. Add the rest of the ingredients, including the Citrus Pursha pulp.
Bring to the boil, add the peel and reduce to a simmer. Once well reduced and candied, remove the peel, drain the syrup into a bowl and set aside, then blend the mix to a paste.
Press through a fine-mesh sieve.
Put the peel in the syrup.
Transfer the paste to a piping bag.

SHORTCRUST PASTRY
Place all ingredients except the eggs in a mixer. Combine roughly using a flat-bladed pastry beater. Add the eggs, then combine again, taking care not to overwork the dough. Roll out to a thickness of 2.5 mm and place in the refrigerator to chill.
Brush the pastry with pure egg yolk, shape into ring doughnuts, and bake at 190 °C for 6–7 minutes.

TOMATOES
Peel the tomatoes and chop into 3 cm-thick slices.
Remove the hearts using a cookie cutter and dice the rest to a fine tartare.
Mix the tartare with the diced Pursha peel.
Season with fine sea salt and leave to drain.

TO SERVE
Place the shortcrust pastry on the plate, add the Pursha paste inside, and top with tomato tartare. Drizzle with half a teaspoon of olive oil and arrange the fennel shoots and flowers around the pastry.
Just prior to serving, heat a frying pan to smoking point and sear the tomato hearts seasoned with a dab of olive oil on each side. Glaze with the Pursha syrup, add a pinch of fleur de sel and place carefully on the tartare.
Add an ice cream quenelle.

SLOW-COOKED PEAS AND WILD MARJORAM

Tender vineyard herbs,
mint and silver fir

SERVES 10

FIR-INFUSED BUTTER

100 g AOP butter
10 g silver fir needles
2 g salt

CREAMED PEAS

100 g minced white onion
300 g fresh peas
300 g roughly chopped peas in their pods
200 g frozen peas
Lemon liqueur, as needed
Lemon-infused olive oil, as needed
Ladle of vegetable stock

SWEATED PEAS

200 g small, tender peas
Pea pod jus
50 g fir-infused butter
Elderflower vinegar, as needed
Ascorbic acid

TO SERVE

Silver fir needles
Wild marjoram
Tender hart's pennyroyal leaves
New grape blossom
Tender grapevine shoots

FIR-INFUSED BUTTER

Soften the butter and blend with the fir needles and salt. Leave to infuse overnight in the refrigerator, then press through a sieve.

CREAMED PEAS

Flash-fry the onions and peas in a lidded pan, adding a splash of oil to set the chlorophyll. Add a ladle of vegetable stock and cook through fast. Blend in a Thermomix and strain through a fine-mesh sieve into a bowl set over ice.
When reheating, loosen with the lemon liqueur and lemon-infused olive oil.

SWEATED PEAS

To extract the juice from the pea pods, take the pods and put them through a juicer. Filter the jus and add a small pinch of ascorbic acid. Gently fry the peas in infused butter with a touch of salt for a few seconds, then gradually add the pod jus. Add fir-infused butter to finish.
Serve in a small bowl as a side dish.

TO SERVE

Grind the fir needles to a very fine sieved powder.
Finely chop the herbs.
Place the creamed peas in the centre of the plate, sprinkle with the finely chopped herbs, then add the various flowers and shoots, emphasising the marjoram and mint.
Dust with the finely ground silver fir.

VENISON LOIN SMOKED WITH HARTWORT ROOT

Marinated in hartwort seeds, barberry vinegar and tender red Swiss chard, and rich pepper caviar jus

SERVES 8

VENISON LOIN
800 g venison loin
10 g dried hartwort seeds
40 g coarse sea salt
5 g dried hartwort root
5 g beechwood sawdust
Lightly salted butter, as needed
Sunflower oil

CELERIAC PUREE
100 g celeriac
200 ml milk

BARBERRY VINEGAR
50 g barberries
50 ml rice vinegar
10 g honey
100 ml water

GARNISH
1 head of red Swiss chard
Fresh barberries

TO SERVE
Hartwort leaves
Fresh butter
Red long peppercorns

VENISON LOIN
Prepare the venison loin, removing the nerves and silverskin.
Break the carcass bones. Roast with the trimmings in sunflower oil and brown off with fresh unsalted butter. Drain off the excess fat, cover with water and leave to simmer for around 2 hours. Filter the jus and reduce to the desired consistency and flavour, regularly skimming off the excess fat.
Grind the hartwort seeds with a mortar and pestle and mix with the coarse sea salt.
Sprinkle the mix over the loin and leave for 4 hours. Smoke the loin with a smoking gun using a blend of crushed roots and sawdust.

CELERIAC PUREE
Peel and dice the celeriac. Cook the chunks in milk.
When tender, strain off the milk and blend the celeriac, then press through a fine-mesh sieve.

BARBERRY VINEGAR
Bring all the ingredients to the boil. Set aside to infuse for 10 minutes before blending thoroughly. Strain through a fine-mesh sieve and reduce to a syrupy texture.

GARNISH
Clean the tender red Swiss chard stems.
Boil the barberries in salt water for 30 seconds, then set aside in the barberry vinegar. Leave to infuse for 7 days, if time permits.

TO SERVE
Finely chop the hartwort leaves.
Roast the loin in a pan with a dab of butter.
Reduce the jus and add a few vinegar-infused barberries and peppercorns.
Place the celeriac puree in a piping bag. Flash-fry a handful of barberries with the Swiss chard leaves.
Set the jus aside in a small saucepan.
Arrange the chopped herbs and other elements decoratively on the plate.

YOUNG CARROTS, FROM LEAF TO SEED

Concentrated toasted seed jus, almost malted reduction
Sabayon infused with vanilla pod, marc de Bourgogne and barrel-aged vinegar

SERVES 4

MINI CARROTS

4 purple carrots
4 orange carrots
4 white carrots
4 yellow carrots
50 ml carrot juice
30 g AOP butter

CARROT SEED MACERATION

5 g carrot seeds
1 spent vanilla pod
1 g tonka bean
10 g honey
100 ml marc de Bourgogne

CARROT PEEL PRALINE

Carrot peelings (reserved)
20 g hazelnuts
Neutral oil

SWEET VERNAL GRASS BUTTER

15 g sweet vernal grass
250 g lightly salted butter

MEYER LEMON GEL

3 Meyer lemons
3 g agar-agar

CARROT REDUCTION

1 kg fresh carrots

SABAYON

40 ml carrot seed maceration
15 g carrot peel praline
110 g egg yolk
Zest of ½ a lemon
133 g sweet vernal grass butter
10 ml aged vinegar
5 ml lemon juice

TO SERVE

Carrot reduction
Marc de Bourgogne caramel
Carrot greens
Fleur de sel
Carrot peel snaps

MINI CARROTS

Peel the carrots, setting the peel aside.
Cook the carrots in the carrot juice. Bring to the boil and add the butter. Season to taste.

CARROT SEED MACERATION

Toast the carrot seeds, then add the honey and sliced vanilla pod. Remove from the heat, then add the grated tonka and marc de Bourgogne. Ideally, leave to macerate for 30 days.
Filter three-quarters of the mix and reduce to a marc caramel.

CARROT PEEL PRALINE

Fry the carrot peelings in a neutral oil at 130 °C. Place on kitchen towel to drain when crunchy. Set some aside for decoration.
Blend the rest with the hazelnuts.

SWEET VERNAL GRASS BUTTER

Chop the sweet vernal grass roughly and heat with the butter over a low flame until all the whey has evaporated.
Strain the warm and browned butter, pressing thoroughly.

MEYER LEMON GEL

Zest half the lemons and squeeze around 100 g of juice. Bring the zests and juice to the boil and add the agar-agar.
Simmer for 1 minute then place in the refrigerator. When the gel is set, blend with a stick blender. Transfer the gel to a piping bag.

CARROT REDUCTION

Juice the carrots in a centrifugal juicer. Filter the juice and caramelise over a hot flame. Transfer to a clean dish and set aside.

SABAYON

Whisk the egg yolks and unfiltered carrot seed maceration into a sabayon over low heat, whisking continuously until thoroughly cooked. Once silky, add the sweet vernal grass butter followed by the rest of the ingredients. Strain thoroughly to filter, season to taste and keep warm.

TO SERVE

Arrange the piping-hot carrots on a plate, add the lemon gel. Sprinkle with the carrot tops and crispy carrot peel. Add the reduction and the caramel, and serve with a generous side serving of sabayon.

CANDIED, CONCENTRATED BEETROOT AND JUNIPER TWIG

Intense pink peppercorn broth, sumac, sorrel and tangy shoots

SERVES 10

BEETROOT
10 small beetroots
Neutral oil
Finely ground juniper
Lime juice and zest
Fine sea salt

JUNIPER VINEGAR
Juniper berries on the twig
Rosso Lambruscum vinegar (Fernando Pensato)
Orange zest

FLAVOURED BUTTER
100 g unsalted butter
1 g sumac
1 g ground juniper
1 g Manakara pink peppercorns
1 cinnamon bud
3 g fine sea salt

SAUCE
200 ml centrifuged beetroot juice
120 g flavoured butter
4 ml gin
Juniper vinegar and a handful of infused juniper berries
Fine sea salt

GIN GEL
100 ml gin
2 g agar-agar

BEETROOT GLAZE
500 ml beetroot juice
100 g fresh butter
Juniper vinegar
Fine sea salt

GARNISH AND TO SERVE
2 bunches of fresh green sorrel
1 yellow beetroot
1 lime
Fleur de sel

BEETROOT
Wrap the individual beetroots tightly in foil with the oil, ground juniper, a quarter of a lime's zest and juice, and salt.
Add another layer of foil and cook in a dry oven (100% humidity, closed vent) at 150 °C for 8–15 hours. Test by inserting a skewer into the centre: it should meet no resistance, and the beetroot should slide off the skewer on its own.
Leave to cool in the foil at room temperature.
Peel with a knife, shaping the beetroot as you go, then halve.

JUNIPER VINEGAR
Collect the juniper berries with as little stem material as possible and fill a 1 litre jar completely, pressing down lightly.
Add the orange zest.
Bring the Lambruscum vinegar to the boil and pour into the jar. If possible, leave to macerate for 1 week.
Dry the leftover twigs in the oven, strip away the thorns and grind to a fine powder. Store at room temperature.

FLAVOURED BUTTER
Soften the butter to a pomade consistency.
Blend the spices to a fine powder and mix into the butter. Set aside to infuse for 24 hours, then divide into 60 g discs. Store in the freezer.

SAUCE
Bring the beetroot juice to the boil, add the rest of the ingredients cold and blend.
Set aside to infuse for 30 minutes, then strain. Adjust the seasoning to taste. Serve on the side.

GIN GEL
Boil the gin and agar-agar, then leave to cool.
Once solid, blend thoroughly and transfer to a pipette.

BEETROOT GLAZE
Heat the cooked beetroots in the beetroot juice with a little salt and vinegar.
Once reduced, add the butter and keep reducing until thick and shiny.

GARNISH AND TO SERVE
Shred the sorrel, cut the yellow beetroot into a fine brunoise and set aside in the juice of one lime.
Arrange the brunoise and gin gel on the plate, add the beetroot and glaze and sprinkle with the shredded sorrel.
Dust lightly with fleur de sel shortly before serving.

LINE-CAUGHT WHITING, ROSEHIP VELVET AND BURGUNDY TRUFFLE

Salsify puree and lemon gel

SERVES 10

WHITING
3 line-caught whiting (400–600 g)
40 g dried kombu
Lightly salted butter
Fine sea salt

ROSEHIPS
1 kg rosehips
100 ml lemon-infused oil
Fish stock
Quince vinegar
Fine sea salt

TRUFFLES
100 g truffles

SALSIFY PUREE
6 salsifies
250 ml milk
250 ml cream
Regular lemon

MEYER LEMON GEL
100 ml Meyer lemon juice
1 g agar-agar
Meyer lemon peel

TO SERVE
Truffle shavings
Olive oil

WHITING
Gut and fillet the fish. Strip out any remaining bones under slow-running cold water, removing the bloodline, fins and gills.
Place the bones and kombu in a large volume of water and boil from cold. Simmer for 40 minutes, skimming as required, then filter through a fine-mesh sieve.
Lightly dust the fillets with fine sea salt and set aside to dry for 4 hours on an oven shelf or clean tea towel. Split the fish down the centre on the skin side and fold it over on itself.
Just before serving, sear the fillets skin-side down in the butter and cook to the desired doneness.

ROSEHIPS
Press the rosehips through a very fine-mesh sieve and set the pulp aside.
Add the remaining solids to the fish stock and simmer for 30 minutes, then press thoroughly through a fine-mesh strainer. Reduce to thicken.
To finish the sauce, stir in the lemon-infused oil and rosehip pulp and season to taste with salt and vinegar.

TRUFFLES
Clean the truffle and cut into generous slices. Set some aside and chop the rest finely. Set aside.

SALSIFY PUREE
Peel the salsifies and chop finely. Set aside in water with a squeeze of lemon juice, topped up with milk and cream to cover them.
Simmer for 20 minutes. Drain and blend thoroughly, adding a splash of the cooking liquid to loosen if necessary. Add salt to taste and set aside.

MEYER LEMON GEL
Put the lemon juice in a pan with the agar-agar. Blend thoroughly before heating, then add the Meyer lemon peel, thinly sliced beforehand. Bring to a simmer and cook for 1 minute.
Place in the refrigerator to cool. Once chilled, blend to a smooth gel, strain through a fine-mesh sieve and transfer to a pipette.

TO SERVE
Lay the fish on a bed of truffle shavings, then decorate the plate with the rosehip puree.
Serve the salsify puree and roseships sauce on the side, topped with the finely chopped truffle, a drizzle of olive oil, and beads of the Meyer lemon gel.

PINK GRAPEFRUIT, POLYPODY FERN AND SAGE

THE DESSERT

by SYLVAIN GOUJON

Polypody fern ice cream, sage crémeux
and Wayanad black pepper

SERVES 10

WHITE SPONGE

45 g whole egg
40 g caster sugar (1)
2 g fine sea salt
12.5 ml grape seed oil
50 g T55 flour
1.5 g baking powder
65 g egg white
17 g caster sugar (2)

GRAPEFRUIT ZEST PICKLES

50 g grapefruit zest
50 ml white rice vinegar
25 g caster sugar
25 ml grapefruit juice

POLYPODY FERN ICE CREAM

210 ml whole milk
90 ml whipping cream
30 g polypody root
5 g Nama sage
35 g caster sugar
20 g glucose powder
20 g skimmed milk powder
5 g ginger

SWISS MERINGUE WITH WAYANAD BLACK PEPPER

50 g egg white
100 g caster sugar
1 g finely ground Wayanad black pepper

CANDIED PINK GRAPEFRUIT WITH PEPPERCORNS

100 g pink grapefruit
60 g sugar (1)
60 ml pink grapefruit juice
7.5 g Madagascar vanilla pod
45 g sugar (2)
12.5 g pectin NH
4 g Wayanad black pepper

POLYPODY FERN AND SAGE CRÉMEUX

260 ml whole milk
30 g polypody root
12 g Nama sage
27 g egg yolk
60 g sugar
7 g cornflour
7 g potato starch
16 g 200-bloom gelatine mass
90 g unsalted butter
Nama sage shoots

WHITE SPONGE

Whisk the egg whites with the caster sugar (1) to stiff peaks.
In a separate bowl, combine the whole eggs, caster sugar (2), salt and oil without whisking.
Sift in the flour and baking powder then fold in the whisked egg whites.
Pour 225 g of batter into a flexible 20 × 30 cm Flexipan tray and place in a fan oven preheated to 180 °C. Immediately lower the temperature to 150 °C and bake for 7–8 minutes.

GRAPEFRUIT ZEST PICKLES

Blanch the grapefruit zest five times, starting with cold water each time.
Reduce the grapefruit juice and sugar to a syrup, then add the white rice vinegar.
Sous-vide the zest and juice and leave to pickle for 1 day before use.

POLYPODY FERN ICE CREAM

Boil the milk in a pan. Add the polypody root and finely chopped sage, set aside to infuse for 20 minutes, strain.
Add the caster sugar, milk powder and glucose powder. Pasteurise at 85 °C. Add the chopped ginger and blend.
Chill rapidly to 4° C and leave to infuse in the refrigerator for at least 4 hours.
Blend and churn in an ice cream maker.

SWISS MERINGUE WITH WAYANAD BLACK PEPPER

Heat the sugar and egg whites in a bain-marie at 65 °C, then whisk to stiff peaks.
Fold in the powdered peppercorns.
Spread on crumpled foil and dry in a cool oven at 60 °C.

CANDIED PINK GRAPEFRUIT WITH PEPPERCORNS

Prick and blanch the pink grapefruit in plenty of water five times, starting with cold water each time.
Slice the grapefruit, removing the central pith and pips.
Blend the vanilla seeds and the sugar (1) in the Thermomix.
Add to a pan with the grapefruit juice and the remaining sugar (2) and pectin, and heat to 50 °C.
Bring to the boil for 1 minute, then add the ground peppercorns.
Chill, blend again, then set aside.

POLYPODY FERN AND SAGE CRÉMEUX

Bring the milk to the boil, then infuse the sage and previously finely sliced polypody root for 20 minutes.
Strain through a fine-mesh sieve, weigh, and add more milk to bring the weight up if necessary.
Combine the sugar, potato starch, and cornflour and whisk with the egg yolks until pale.
Bring to the boil with the infused milk, then add the gelatine mass. Chill to 40 °C, add the softened butter, then blend to a smooth texture.
Chill and set aside.

TO SERVE

Brush the white sponge generously with grapefruit zest pickle juice.
Cut into 6 × 3 cm rectangles.
Pipe on a 2 cm-thick layer of polypody fern ice cream.
Spread the candied grapefruit on the plate. Pipe on beads of sage crémeux and decorate with quarter-slices of grapefruit, grapefruit zest pickles and Nama sage shoots.
Top with the ice cream and meringue.

ACKNOWLEDGEMENTS FROM BRICE DE LA MORANDIÈRE

The pleasure of wine is a symphony, performed by an orchestra we hope is talented, based on an original score – the terroir – with instruments that the musicians have tended with the utmost care: their vineyard. Every musician contributes to the final harmony, from the conductor to the first violin, down to the humblest of instruments – sometimes a simple triangle.

Thank you to the team behind *A Tribute to Life*: Jean-François Aimé, Garlone Bardel, Hugo Bourny, Jérôme Bryon, Valérie Fuchs, India Muël, Guillaume Rebière and Thierry Thévenin.

Thank you to the team at Domaine Leflaive who worked specifically on this project: Laurent Alexandre, Justine Belloc, Julie Boisselet, Louis Brière, Amandine Brillanceau, Bertrand Chapuis, Julien Coyaud, Carole Dupaty, Baptiste Guigon, Bastien Muzard and Esteban Pavesi.

Thank you to all the other talented contributors: Caroline Besse, Anaïs Bourny Delon, Henry de Monclin, Carole Dumont, Gabriel Faquet, Sylvain Goujon, Valérie de La Morandière, Julien Morvan, Martin Muriot, Philippe Rodriguez, Anna Saint Pierre, Marco Simonit, Victoria Sosolič and Françoise Vannier.

Left
Brice de La Morandière

Right
Common ball flower

DOMAINE LEFLAIVE

Edited by
Brice de La Morandière

Project manager and editorial designer
Valérie Fuchs

Art director
Jean-François Aimé

Graphic design and layout
India Muël et Jean-François Aimé

Photographer
Jérôme Bryon

Author
Brice de La Morandière

Author
Guillaume Rebière

Botanist, forager, author
Thierry Thévenin

Michelin-starred chef at Lucas Carton
Hugo Bourny

Pastry chef
Sylvain Goujon

Culinary stylist
Garlone Bardel

ÉDITIONS SKIRA PARIS
7 rue Kepler
75116 Paris
www.skira-arte.com

Senior editor
Nathalie Prat-Couadau

Editorial coordinator
Victoria Calligaro

Editorial manager
Maria Notó Mora

Editor
Roxanne Rebours

Editorial assistant
Lou Brunet (intern)

Business development officer
Bertille Bornens (intern)

Art director
Diane de Noyelle

Layout
Mathis Bécard

Translation
French to English Ella Fox Plaisted and Susan Pickford

Copy-editing
English Myriam Birch

Photo credits
© Jérôme Bryon
Except pp. 10, 33, 198 © Roberta Valerio
p. 213 © Domaine Leflaive

Colour separation
Litho Art New, Turin

ISBN 978-2-37074-311-4

Printed in July 2026 by Graphius, Ghent, Belgium
Legal deposit: August 2026

MIXTE
Papier | Pour une gestion forestière responsable
FSC® C014767

This book has been printed on FSC-certified paper, and all stages of its manufacture have complied with this certification, which supports the environmentally appropriate, socially beneficial and economically viable management of the world's forests, using materials from well-managed forests, recycled materials and other controlled sources. www.fsc.org